INEQUALITY, DEMOCRACY, AND
THE ENVIRONMENT

Inequality, Democracy, and the Environment

Liam Downey

NEW YORK UNIVERSITY PRESS

New York and London

NEW YORK UNIVERSITY PRESS
New York and London
www.nyupress.org

References to Internet websites (URLs) were accurate at the time of writing. Neither the author nor New York University Press is responsible for URLs that may have expired or changed since the manuscript was prepared.

ISBN: 978-1-4798 5072-3 (hardback)
ISBN: 978-1-4798-4379-4 (paperback)

For Library of Congress Cataloging-in-Publication data, please contact the Library of Congress.

New York University Press books are printed on acid-free paper, and their binding materials are chosen for strength and durability. We strive to use environmentally responsible suppliers and materials to the greatest extent possible in publishing our books.

Manufactured in the United States of America

10 9 8 7 6 5 4 3 2 1

Also available as an ebook

This book is dedicated to all those who are struggling to make the world a more democratic, just, and sustainable place.

CONTENTS

ACKNOWLEDGMENTS

First and foremost, I would like to thank the graduate students who helped me conduct much of the research for this book: Susan Strife, Brian Hawkins, Katherine Clark, and Eric Bonds. Without their help this book would not have been written, and I am especially indebted to Katherine and Eric for their assistance, friendship, and support. I would also like to thank my friends and colleagues at the University of Colorado–Boulder, in particular, Jenn Bair and Isaac Reed, who provided helpful advice and support as I developed my initial ideas into a more formal theoretical model and then into this book.

I have taught the material covered in this book to hundreds of graduate and undergraduate students over the past 10 years, many of whom have provided me with important feedback on the ideas included in this book. They are too many to list here, but I thank all of them. In addition, there are several people outside academia whom I would like to thank. The friends I have made in Boulder's music scene have, simply by being there week in and week out, made the writing of this book much easier. I would like to thank all of them, but in particular Mark Arroyo, Ray Alonso, and Jodie Woodward. My wife, Alessandra, and sons, Naiche, Dakota, and Maja, have also supported me throughout the writing of this book. Alessandra also read and provided important feedback on the final draft of my manuscript. Thank you.

Finally, I cannot express how grateful I am for the enthusiastic support and advice of Ilene Kalish, Caelyn Cobb, and Dorothea Halliday, executive editor, assistant editor, and managing editor, respectively, at NYU Press. My book also benefited greatly from Andrew Katz's careful editing, and I am extremely grateful for the feedback provided by the scholars who anonymously reviewed my manuscript. This book is much better for their advice and suggestions.

Introduction

The world currently faces several severe social and environmental crises, including economic underdevelopment, widespread poverty and hunger, lack of safe drinking water for one-sixth of the world's population, deforestation, desertification, rapidly increasing levels of pollution and waste, dramatic declines in biodiversity and soil fertility, and climate change. Evidence of these and other serious social and environmental problems abounds. For instance, well over two billion developing nation citizens live on less than two dollars a day (Shah 2010a), there are currently one billion people around the world who do not have enough food to lead healthy, active lives (FAO 2009), more than 24,000 children die every day due mainly to poverty, hunger, and preventable diseases (Shah 2010b), more than 90 million tons of greenhouse gases are spewed into the atmosphere each day (Cunningham and Cunningham 2007; Gore 2009; IPCC 2007; Lappe 2010), and each year five million acres of rainforest are cut down, five to six million people die prematurely from exposure to air pollution, and roughly 75 billion metric tons of topsoil are lost to erosion (Cunningham and Cunningham 2007).

A major source of many of these social and environmental problems is modern agriculture, which routinely poisons farm workers, generates unbelievably large volumes of animal waste (more than half a billion tons of waste per year in the U.S. alone), places highly toxic residues in our food, makes the world's food supply highly susceptible to pests, disease, and climate change, produces widespread poverty and hunger, poisons and depletes freshwater sources around the world, and contributes greatly to deforestation, soil erosion, and climate change (Cunningham and Cunningham 2007; Hawken et al. 1999; also see chapter 4).

The extraction of natural resources also severely harms people and the environment. Over the past 50 years, for example, an estimated 540 million gallons of oil have been spilled or otherwise released into Nigeria's highly sensitive delta region (Nossiter 2010), devastating the local environment and resulting in the widespread loss of livelihoods for people in the region (J. Brown 2006). Nigerian oil production has also sparked armed violence and repression as local people, rebel groups, and Nigerian and oil-company military personnel have struggled for control over the region's oil and local environment (J. Brown 2006; Human Rights Watch 1999; Watts 2008).

Of course, Nigeria's delta region is not the only part of the world that has been devastated by the extraction of natural resources. Other examples of highly destructive resource extraction activities include British Petroleum's Deepwater Horizon oil spill, which released 205.8 million gallons of oil into the Gulf of Mexico (Robertson and Krauss 2010), Texaco's deliberate release of 345 million gallons of crude oil and 16–18 billion gallons of toxic waste into the Ecuadorian rainforest between 1964 and 1990 (Environment News Service 2012; Rainforest Action Network 2012; *Scandinavian Oil Gas Magazine* 2010), the removal of mountain tops and destruction of local waterways by mining companies in Appalachia to more cheaply exploit the region's coal reserves (Scott 2010), and the generation of 200,000 tons of mine waste per day at the Grasberg copper mine in West Papua, Indonesia (Downey et al. 2010). Moreover, wars are often fought to gain or maintain control over natural resources, resulting in widespread death, severe social dislocation, and further environmental destruction (Klare 2001).

It goes without saying that these social and environmental problems are all extremely serious and that they all demand immediate attention. The question is, how should we go about trying to address them? Are these problems each the product of a unique set of social forces such that each problem must be dealt with separately? Or is there a factor or set of factors common to all of them that must be addressed before any of them can be solved? And if there is a factor or set of factors common to all of them, what could this factor or these factors be?

Many different answers have been given to these questions. For instance, many mainstream environmentalists believe that social and environmental problems arise from fundamentally different causes, that some problems, in particular global climate change, are so serious that they must be solved before we devote time and resources to addressing other problems and that the best way to solve or ameliorate specific environmental problems is to ignore broad social forces that likely play a key role in creating these problems and instead craft environmental agreements, legislation, or regulations that only address one problem, or a narrowly defined set of problems, at a time.[1] These mainstream environmentalists thus focus their efforts on specific laws, regulations, treaties, organizations, and behaviors, working on one or two environmental issues at a time and ignoring the broader social structures and power relations within which individuals and organizations operate and within which laws, regulations, and treaties are legislated, negotiated, promulgated, and enforced.

In contrast, many other mainstream environmentalists argue that there *is* a single factor driving the environmental crisis—environmentally harmful consumer behavior. These environmentalists thus contend that the solution to the crisis is for consumers to voluntarily adopt pro-environmental purchasing behaviors, which, in turn, will promote pro-environmental business practices.

Another group of environmentalists, environmental economists, argue that environmental problems result from a combination of poorly functioning and structured markets and excessive government interference in the market, which they contend promote environmentally harmful behavior by distorting the price signals that consumers and businesses receive. They therefore conclude that in order to solve the world's critical environmental problems, we must reduce government interference in local, national, and global markets and properly structure these markets.

Finally, a group of environmental sociologists that I call macro-structural environmental sociologists tend to place blame for social and environmental problems on local, national, and global social structures (see chapter 2 for a definition of the term *social structure*), arguing for example that these problems are the product of factors and forces such as structured inequality, the exchange relations inherent in global trade, a treadmill of production that forces capitalist firms to continually expand production, and the constraints placed on world leaders by the dynamics of the capitalist world system.

Proponents of these four perspectives thus provide very different explanations of and very different solutions for the global environmental crisis, with some mainstream environmentalists exhorting individuals to change their consumption practices; other mainstream environmentalists taking an incremental approach that seeks to create, abolish, or amend, in a piecemeal fashion, specific rules, laws, regulations, and treaties; still other mainstream environmentalists (environmental economists) promoting market reforms designed to alter individual and business behavior; and macro-structural environmental sociologists advocating broad changes to local, national, and global social structures that would drastically alter capitalist social relations, greatly reduce local, national, and global inequality, fundamentally reshape the way that markets function, and greatly alter the incentive structures that shape individual, group, and business behavior.

It clearly matters, then, whether incremental environmentalists are correct in believing that the world's myriad social and environmental problems arise from fundamentally different causes such that specific environmental problems can be dealt with separately from other social and environmental problems, whether environmental economists are correct in identifying poorly functioning markets and overreaching governments as the ultimate source of the world's environmental problems, whether those who exhort consumers to change their purchasing habits are correct in placing primary blame for the environmental crisis on consumers, or whether macro-structural environmental sociologists are correct in highlighting local, national, and global social structures as the key obstacles to achieving social justice and environmental sustainability. In addition, because proponents of these perspectives all believe that individual and business behavior must change if we are to save the environment, it is also important to determine which of these approaches will best promote behavioral change among individuals and businesses. Is it enough to simply exhort consumers to change their behaviors or to create laws, treaties, and regulations for specific environmental problems? Or do we need to greatly alter the incentive structures that individuals and businesses face by implementing far-reaching market reforms or drastically changing the social structures that most profoundly shape our lives?

As a macro-structural environmental sociologist, I believe the evidence strongly supports the argument that local, national, and global social structures are largely to blame for the world's many serious social and environmental problems. Accordingly, much of this book is devoted to demonstrating that this is the case and that market reforms, piecemeal legislation, and exhorting individuals and organizations to change their behavior will not solve the world's myriad social and environmental problems. However, unlike other macro-structural environmental sociologists, my central argument in this book is that the world's social and environmental problems are to a significant degree the product of organizational, institutional, and network-based inequality (OINB inequality), which I hold plays a key role in creating and maintaining many of the social structures and forms of structured inequality highlighted by other macro-structural environmental sociologists.

It is true, of course, that macro-structural environmental sociologists focus significant *empirical* attention on many of the organizations, institutions, and networks that I examine in this book, as well as on the inequality inherent in these organizations, institutions, and networks. Nevertheless, it is also true that as a *theoretical* construct, OINB inequality has largely been ignored by macro-structural environmental sociology. This is a critical mistake, I argue, because OINB inequality provides economic, political, military, and ideological elites with the means to shift environmental and non-environmental costs onto others by giving these elites the power to develop and control undemocratic organizations, institutions, and networks through which they are able to achieve socially and environmentally harmful goals in the face of resistance from others. In other words, OINB inequality is theoretically and substantively important, I argue, because it ensures that elites are better able than other social groups to achieve their goals and because the goals that elites want to achieve are often socially and environmentally harmful. (My point here is not that elites necessarily want to harm individuals, societies, and the environment but rather that this is an important result of their actions.)

I further contend that the form that elite-controlled organizations, institutions, and networks take and the specific functions they perform can vary widely over time and from one place to another[2] because elite-controlled organizations, institutions, and networks are embedded in specific economic and political systems and sets of social relations at the local, national, and global levels that vary both geographically and temporally and that shape and are shaped by the institutions, organizations, and networks that constitute them. Thus, my argument is not that OINB inequality and elite-controlled organizations, institutions, and networks shape social and environmental outcomes independently of more general factors such as the logic of capital accumulation, the dynamics of the capitalist world system, the modes and relations of production, or national political and ideological systems, but rather that elite-controlled organizations, institutions, and networks are the concrete tools that elites use to achieve goals such as accumulating capital, maintaining capitalist social relations, dominating public policy making, and shaping public opinion (M. Mann 1986). They are also the settings within which and through which elites make decisions, determine their interests, and exert power.

As a result, unlike other macro-structural environmental sociologists, who tend to focus theoretical and empirical attention on large-scale social forces such as the expansionary logic of capitalism and the dynamics of the capitalist world system, I focus my theoretical and empirical attention instead on the organizations, institutions, and networks that make capitalist expansion and world-system dynamics possible and that enable elites in any complex society and any economic and political system to achieve their goals in the face of resistance from others.

I therefore contend that despite the wide variety of forms and functions that elite-controlled organizations, institutions, and networks can take, they nevertheless represent an important category of social mechanism that plays a key role in harming individuals, societies, and the environment in all complex societies and world systems. I further contend that they play a theoretically similar role in producing social and environmental harm wherever and whenever they are found; that they are necessarily undemocratic because elites are greatly outnumbered by non-elites and, as a result, cannot afford to provide non-elites with democratic access to the organizations, institutions, and networks that elites wish to control; and that they tend to be violent in at least one of three ways: they either (a) use or threaten to use armed violence to achieve their goals, (b) achieve their goals by relying at least in part on armed violence carried out by other organizations, institutions, or networks, or (c) severely harm, either physically, emotionally, or psychologically, individuals, communities, societies, and/or the natural world (see chapter 5).

For example, the World Bank and agricultural commodity chains are very different kinds of institutions with goals that partially overlap but are also quite distinct. Nevertheless, as I discuss in detail later in the book, they are both undemocratic and controlled by elites; they both provide elites with the means to achieve their goals in the face of resistance from others, in large part by allowing elites to monopolize decision making within them; and because achieving the goals elites set for them often results in environmental degradation and social dislocation, they both harm individuals, societies, and the environment in theoretically similar ways. One way the World Bank does this is by forcing developing nations to support the existence and creation of large, foreign-owned mines that wreak havoc on local communities and environments but help the World Bank achieve its goals of ensuring a secure supply of raw materials for industry and opening developing countries to foreign investment (see chapter 5). Lead firms in agricultural commodity chains do this, at least in part, by using their power within these chains to dictate to farmers and ranchers the agricultural production practices they must use, practices that increase corporate profits, produce poverty and hunger for farmers around the world, and harm the environment by relying on petroleum-based pesticides and herbicides, heavy machinery, and extraordinarily large "confined animal feeding operations" that concentrate livestock and their waste in relatively small areas (see chapter 4).

So while on the surface it may seem as though the World Bank and agricultural commodity chains have little in common with each other, a careful comparison of these two institutions shows that they both contribute to the global environmental crisis in very similar ways. They are both powerful social mechanisms that provide specific groups of elites with the means to monopolize decision making power, shift environmental and non-environmental costs onto others, and limit the pro-environmental behavior of individuals and governments. As a result, these institutions provide those elites who dominate them with the power to achieve goals that, intentionally or not, harm individuals, communities, and societies and degrade the environment. Moreover, because the social and environmental harm these institutions produce is so severe, they are both extremely violent in their effects.

In order to demonstrate the validity and widespread applicability of my theoretical argument, the empirical chapters in this book present a series of case studies that highlight several of the United States' and the world's most important elite-controlled organizations, institutions, and networks and show that these organizations, institutions, and networks play a key role in shaping some of the world's most critical human, social, and environmental crises. These case studies cover a wide variety of social and environmental topics, allowing me to demonstrate that undemocratic and elite-controlled organizations, institutions, and networks as diverse as the World Bank, commodity chains, policy planning networks, and the news media belong to a single category of social mechanism that is responsible for much of the social and environmental devastation the world currently experiences and that operates in theoretically similar ways in a wide variety of substantive settings. The broad social and environmental scope of these case studies also allows me to demonstrate that researchers and activists cannot fully explain the structural causes of environmental degradation without examining either the link between OINB inequality and environmental degradation or the organizational, institutional, and network-based mechanisms through which elites exert power.

In providing strong empirical support for my theoretical argument, these case studies also help me to advance macro-structural environmental sociology both substantively and theoretically. In particular, by including undemocratic and elite-controlled organizations, institutions, and networks within the subfield's theoretical purview and by demonstrating to a nonsociological audience the important role that structural forces play in shaping social and environmental outcomes, this book fills important theoretical and substantive gaps in the subfield while simultaneously helping the subfield achieve its goal of convincing environmentalists and others to pay more attention to elite-dominated social structures.

Moreover, in demonstrating that inequality, power, weak democratic institutions, and local, national, and global social structures play a fundamental role in producing severe social and environmental devastation, the case studies presented in this book also demonstrate that incremental approaches to solving environmental and social problems will not work and that consumer behavior and poorly functioning markets are not the root causes of these problems. This implies, of course, that attempts to convince consumers to change their purchasing behavior and efforts to pressure governments to sign treaties, reform their markets, and draft new laws and regulations will not, in and of themselves, solve the world's myriad social and environmental problems.

Finally, and perhaps most importantly, by highlighting a set of elite-controlled organizations, institutions, and networks that play a critical role in producing social and environmental harm, the case studies included in this book help identify important elements of the social structure that must be radically altered or abolished if activists are to achieve social justice and environmental sustainability.

The book is divided into two sections. In the first section, I set up and present my theoretical argument. Thus, in chapter 1 I summarize and critique two of the most popular explanations of environmental degradation currently found in environmental and public-policy circles. The first explanation I critique is the mainstream environmentalist argument that consumers are the main drivers of the global environmental crisis. I then critique environmental economics, which holds that poorly functioning markets and overreaching governments are primarily to blame for this crisis. After critiquing these two arguments, I note that one problem they both share is that they tend to downplay, ignore, or inadequately theorize inequality, power, and democracy. I explain why this is problematic and argue that the theoretical model I develop and evaluate in this book is designed in part to overcome the shortcomings of these two explanations of the environmental crisis.

In chapter 2 I present my theoretical argument. I begin by defining the terms *social structure* and *macro-structural environmental sociology* and then discuss the strengths and weaknesses of the two most prominent theoretical approaches found within macro-structural environmental sociology. I use this discussion to identify important gaps in the macro-structural environmental sociology literature and argue that my theoretical model fills these gaps in several important ways. I then lay out my theoretical argument, drawing primarily on theory and research from environmental and political sociology.

In the second half of the book I present my empirical argument. I focus substantial attention in this part of the book on U.S.-dominated global institutions and the effects that these institutions and elite-controlled organizations, institutions, and networks in the U.S. have on social and environmental conditions in the U.S. and around the world. In addition, I limit my investigation to events that have occurred in the past 70 years, with most of my investigation restricted to the past 30–40 years. There are two main reasons for my focus on U.S.-based organizations, institutions, and networks and U.S.-dominated global institutions in the post–World War II period. First, if we are to solve the world's social and environmental crises, we must identify those elite-controlled organizations, institutions, and networks that *currently* shape these crises because it is these organizations, institutions, and networks that must be altered or abolished if we are to create positive social and environmental change. Second, despite the apparent decline in power of the United States over the past 50 years (Wallerstein 2003), the U.S. is still the most economically and militarily powerful nation in the world, as well the nation that consumes the largest share of the world's natural resources (Markham and Steinzor 2006). As a result, understanding how elite-controlled organizations, institutions, and networks in the United States shape social and environmental outcomes both in the U.S. and globally is no trivial matter. Focusing on the U.S. is also important because the U.S. is the dominant actor in the world's most socially and environmentally destructive international institutions, institutions such as the World Bank and the International Monetary Fund (IMF) that feature prominently in this book and that play key roles in shaping social, political, economic, and environmental outcomes around the world.

Thus, in focusing much of my attention on U.S.-based organizations, institutions, and networks and U.S.-dominated global institutions in the post–World War II period, I am not repudiating my argument that OINB inequality and elite-controlled organizations, institutions, and networks play a key role in shaping social and environmental outcomes in *all* complex societies and world systems. Rather, my focus on the U.S. and the post–World War II period reflects both my desire to understand and solve current social and environmental problems and the fact that since World War II, the U.S. and U.S. elites have been the most powerful nation and group of elites in the world, a position they have maintained in large part through their domination of international institutions such as the World Bank and IMF.

I have six overarching goals in the second half of the book: to demonstrate that OINB inequality and undemocratic and elite-controlled organizations, institutions, and networks play a key role in harming individuals, societies, and the environment; to highlight a variety of organizational, institutional, and network-based mechanisms that elites use to achieve their goals; to show that these mechanisms operate in theoretically similar ways regardless of their form or specific function; to demonstrate the empirical validity of my theoretical model by (a) testing a series of hypotheses that I derive from the model and (b) showing that the model works in a variety of substantive settings; to demonstrate that elite-controlled organizations, institutions, and networks are inherently violent; and to identify places in the social structure where drastic corrective action is needed to begin solving the world's myriad social and environmental crises.

To achieve these goals, I present a series of case studies that on the one hand examine four substantive topics—globalization, agriculture, mining, and U.S. energy and military policy—and on the other hand highlight a broad range of elite-controlled mechanisms that have produced severe social and environmental harm. These mechanisms are all rooted in elite-dominated organizations, institutions, and networks and include armed violence, oligopoly and oligopsony power, free trade agreements, property rights protections, policy planning networks, political risk insurance, the news media, international debt, structural adjustment, and decision making rules at the World Bank, IMF, and World Trade Organization (WTO). This is clearly a lot of ground to cover, but only by covering so much ground can I demonstrate the fundamental social and environmental significance of elite-controlled organizations, institutions, and networks and the widespread applicability of my theoretical model.

Thus, in chapter 3 I argue that globalization has some very specific social and environmental consequences that are the product not of inevitable and inexorable economic forces but of specific decisions made by powerful actors in a set of elite-controlled institutions that have worked very hard to create a global neoliberal order. After setting forth this argument, I narrow my focus to two of these institutions: the World Bank and the IMF. I explain who runs these institutions and how they operate and demonstrate that not only are they the product of OINB inequality and various elite-controlled organizational mechanisms, they are also important elite-controlled mechanisms in their own right. I then explain how they simultaneously harm individuals, societies, and the environment and provide elites with the means to achieve their goals in the face of resistance from others. Finally, I provide evidence that this nation's paper of record, the *New York Times*, provides readers with little information regarding the negative social and environmental consequences of World Bank policy or the role the U.S. and U.S. elites play in shaping World Bank policy. This, I argue, suggests that the corporate media in the U.S. keep citizens largely uninformed about how U.S.- and elite-dominated international institutions affect people and the environment around the world, thereby minimizing the likelihood that U.S. citizens will resist the goals that elites want these institutions to achieve.

The following chapter, chapter 4, examines eight organizational, institutional, and network-based mechanisms that economic and political elites in the U.S. use to achieve agriculture industry goals: oligopoly and oligopsony power in agricultural commodity chains; liberalized agricultural trade; expansive property rights protections; free trade agreements that promote corporate-friendly property rights protections and liberalized agricultural trade; the organizational links that exist between members of a group of capitalists called the power elite that allow them to mobilize corporate and capitalist support for specific public policies; the organizational links that exist between members of the power elite and the U.S. government that allow the power elite to strongly shape U.S. policy; decision making rules at the WTO that provide the U.S. (and Europe) with overwhelming advantages during WTO negotiations; and economic threats made by the U.S. during WTO negotiations to ensure cooperation from developing nations.

The chapter begins with a discussion of several of the most serious social and environmental problems associated with modern, industrial agriculture. It then links these problems to the structure of the modern agriculture industry and explains how the eight undemocratic and elite-controlled mechanisms highlighted in the chapter have simultaneously structured this global industry, harmed individuals, societies, and the environment, and provided elites with the means to achieve their goals. The chapter demonstrates, for example, that elite-controlled policy planning networks in the U.S. play a key role in shaping U.S. and WTO policy regarding agricultural trade, which, in turn, benefits agribusiness corporations and severely harms individuals, communities, and the environment around the world. The chapter also explains how agribusiness firms and supermarkets use their oligopsony and oligopoly power in agricultural commodity chains to reduce farmer profits, shape farmer behavior, and accumulate capital.

I shift gears in chapter 5 by turning my attention to natural resource extraction and armed violence. I begin by arguing that armed violence is one of several overlapping and mutually reinforcing elite-controlled mechanisms that provide core nations and corporations with the means to control or gain disproportionate access to the natural resource wealth of developing nations (core nations are the world's wealthiest nations). I then demonstrate that armed violence is strongly associated with natural resource extraction in many parts of the world. To demonstrate this association, I identify 10 minerals that are critical to the functioning of the U.S. economy and/or military and ask whether the extraction of these critical minerals involved the use of armed violence at any point between 1995 and 2009 (the research for this portion of chapter 5 was conducted in 2009). I supplement this analysis with a set of short case studies that more fully examine the violent activities associated with the extraction of two of these critical minerals (manganese and copper) and then provide a general discussion of the role that armed violence played in supporting the activities of what in 2007 were the world's three largest mining companies (Rio Tinto, CVRD/Vale, and BHP Billiton). Finally, to demonstrate that armed violence works in conjunction with other elite-controlled organizational mechanisms, I examine several mechanisms that the World Bank uses to promote natural resource extraction in Africa and investigate the degree to which armed violence carried out by military, police, mercenary, and rebel forces is associated with African mining projects that have received World Bank funding or World Bank investment guarantees. I conclude that armed violence is a critically important elite-controlled mechanism that in conjunction with other elite-controlled mechanisms ensures core nations' access to and control over natural resources, thereby promoting capital accumulation and military power in the core, degrading local, regional, and global environments, and creating the conditions within which ecological unequal exchange (see chapter 5) can occur.

In chapter 6, the final empirical chapter of the book, I examine the elite roots of U.S. energy and military policy in the George W. Bush administration. I briefly describe U.S. energy and military policy since World War II, discuss the human, social, and environmental consequences of this policy in the U.S. and around the world, and then examine the role that economic and political elites, the U.S. policy planning network, and restricted decision making played in the formation of the Bush administration's military and energy policy. As in the previous empirical chapters, I highlight the importance of OINB inequality and elite-controlled organizations, institutions, and networks and demonstrate that the empirical evidence presented in the chapter is consistent with the theoretical argument I set forth in chapter 2.

Finally, in the concluding chapter of the book I summarize my theoretical argument and empirical findings and discuss the theoretical and substantive implications of these findings. Most importantly, I argue that the evidence presented in the case studies strongly supports my theoretical model, thereby demonstrating that sociologists and environmentalists cannot explain the structural causes of environmental degradation without examining either the link between OINB inequality and environmental degradation or the organizational, institutional, and network-based mechanisms through which elites exert power.

I turn now to a critique of environmental economics and consumer-oriented environmentalism.

1

Popular Explanations of the Environmental Crisis

On April 20, 2010, the Deepwater Horizon oil rig, leased by the oil giant BP, exploded and sank off the coast of Louisiana, killing 11 and injuring 17. Two days later, oil was discovered leaking from the damaged well site, and by the time the well was capped nearly three months later, an estimated 4.9 million barrels, or 205.8 million gallons, of oil had spewed into the Gulf of Mexico (Robertson and Krauss 2010). One of the worst accidental oil releases ever, the BP oil spill, and the toxic dispersants BP used to break down the spilled oil, continue to threaten thousands of animal and plant species that rely on the Gulf ecosystem for food, habitat, and breeding (Biello 2010; National Wildlife Federation 2015; Iallonard 2010). Of particular concern to scientists are the effects of the spill and dispersants on fish eggs, larvae, crabs, oysters, birds, mammals, fish, insects, and sensitive habitats such as beaches, wetlands, mudflats, estuaries, and corals. There is evidence, for instance, that deep-sea corals have been severely damaged by the spill (Press Association 2012), that contamination from the spill is moving up the food chain from phytoplankton and zooplankton to larger marine organisms (ABC News 2012), that deep-sea shrimp, crab, and lobster are experiencing higher than usual deformity rates (Democracy Now 2012), that red snapper are 20–50 times more likely than before the spill to have lesions and other physical problems (Democracy Now 2012), and that dolphins and sea turtles in the area are dying in larger than normal numbers (ABC News 2012). In addition, immense volumes of oil and dispersants have settled at the bottom of the ocean floor, waiting to be stirred up by the next big tropical storm (ABC News 2012).

Because the Gulf Coast economy relies heavily on tourism and commercial and recreational fishing, which, in turn, depend on the health and beauty of the local ecosystem, the BP oil spill has also damaged the regional economy. For example, prior to the spill (in 2008), the dockside value of commercially harvested fish and seafood in the Gulf Coast region was $659 million, recreational fishing trip expenditures were valued at $1.5 billion, and 25,700 people were employed in the commercial fishing industry. During the same year, approximately $51 billion was spent on tourism in the region, resulting in $17.6 billion in wages for the 587,000 Gulf Coast residents employed in the industry. Thus, the decline in tourism and temporary closure of 88,522 square miles of federal Gulf waters to fishing after the oil spill caused immediate and severe economic hardship for many individuals in the region, as evidenced by the fact that by August 20, 2010, over 150,000 monetary claims, representing approximately 250,000 separate damages, had been filed against BP by Gulf Coast residents (White House Oil Spill Commission 2010). Moreover, in 2012 researchers were still unsure whether the region's tourist industry had fully recovered (Finn 2012), and local fisherman continued to claim that their catches were significantly lower than they had been before the oil spill (Jamail 2012; Johnson et al. 2012).

The spill also poses a serious health risk to those who were exposed to the spilled oil, to fumes from the oil, or to the toxic chemicals used during cleanup efforts (Canfield 2012; White House Oil Spill Commission 2010), as well as to those who have eaten seafood contaminated by the spill and dispersants, a risk that is of particular concern for pregnant women and children (Rotkin-Ellman et al. 2012). Concerns about the long-term social, economic, and medical consequences of the oil spill have also bred anger and resentment among many local residents (Jamail 2011), which, in turn, could result in increased levels of domestic violence, substance abuse, and stress-related illnesses (White House Oil Spill Commission 2010).

Thus, as with most environmental crises, the BP oil spill has had and will continue to have multiple social, economic, and environmental consequences that are difficult to disentangle from one another. As with other environmental crises, the BP oil spill has also prompted many people to offer explanations for why the disaster occurred and to provide prescriptions for how to avoid similar disasters in the future. For instance, a perusal of letters to the editor at the *New York Times* shows that many readers thought the spill was the product of BP's greed and ability to avoid regulatory oversight, while other readers argued that the disaster resulted from high consumer demand for oil, with some blaming consumers (rather than BP) for the oil spill and others suggesting that consumer demand for oil is high because the price of oil is too low given its high social and environmental costs. These letters thus offer several implied or explicitly stated suggestions for how to avoid similar disasters in the future, including subordinating corporate power to regulatory oversight, getting consumers to voluntarily change their purchasing behavior, and incorporating the social and environmental costs of fossil fuel use into the prices we pay for these fuels.

What is striking, though not particularly surprising, about these letters to the editor is how closely they hew to a handful of relatively well-known explanations of and solutions for the global environmental crisis. For example, portraying consumers as both the problem and the potential solution to the environmental crisis is, as I noted in the introduction, a well-known argument among environmentalists, many of whom believe that we can solve the crisis simply by altering our purchasing habits. Similarly, much attention has been devoted in the news media and environmental and public policy circles to the environmental economics argument that the best way to solve environmental problems is to incorporate the financial costs of these problems into the prices that businesses and consumers pay for goods and services. And finally, the liberal and left-leaning press has long argued that many of our nation's social and environmental problems result from a lack of regulatory oversight by government over business.

Given the widespread support these arguments have found among certain segments of the public, the fact that the consumer behavior and environmental economics arguments have eclipsed the argument for environmental regulation in many environmental and public policy circles, and my belief that the consumer behavior and environmental economics arguments are flawed in fundamentally important ways,[1] my goal in this chapter is to explain why I think these two arguments are flawed so that I can focus later in the book on factors that I believe are more central to explaining and solving the world's many social and environmental crises.

It is important to note, however, that in critiquing these two arguments, I do not mean to imply that they provide us with no important insights. Changing consumer behavior and altering the manner in which we determine the prices of goods and services are probably both necessary ingredients for overcoming inequality, injustice, and the world's major social and environmental problems. Nevertheless, as I argue in both this and the next chapter, a number of fundamental social forces and social structures exist that prevent these and other environmentally and socially important changes from occurring. As a result, these forces and structures must be identified, understood, and overcome before we can fundamentally alter consumer behavior, incorporate a broad range of social and environmental costs into prices, and solve the world's major social and environmental crises.

There are also many ways in which consumer behavior and prices can change that, although consistent with the consumer behavior and environmental economics arguments, will not solve the environmental crisis or promote social justice. For example, if we shift our purchasing habits so as to buy less environmentally damaging products, but dramatically increase the number and volume of products we purchase, we are still likely to increase the environmental burden we place on the planet, a dynamic that already exists throughout much of the world (York and Rosa 2003). Similarly, several highly regarded policies for incorporating social and environmental costs into the prices of goods and services have been proposed that, if enacted, are likely to increase economic inequality—by, for example, placing increased economic burdens on the poor and working class and increasing Wall Street profits—while doing little to help the environment (see later in this chapter).

Thus, in the following sections of the chapter, I provide a critique of the consumer behavior and environmental economics arguments that highlights several fundamental flaws in these arguments, one of which is that they ignore inequality, power, and other important social structural factors, a characteristic they share with many mainstream explanations of the environmental crisis. Carefully evaluating these two arguments will therefore tell us much about what is wrong with mainstream environmentalism and why environmentalists must incorporate social structural factors such as OINB inequality and elite-controlled organizations, institutions, and networks into their explanations of the environmental crisis.

I turn now to a critique of the consumer behavior argument.

Consumers as Environmental Saviors

Undoubtedly BP is responsible [for the Gulf Coast oil spill], but so are all of us who drive cars, travel by plane or consume goods produced and shipped with oil. If we didn't use it, BP wouldn't drill for it. Until we recognize that demand for oil is as much the problem as supply, and start to change the way we live to reduce it, environmental destruction is inevitable.

—We are all BP (M. Brown 2010)

If you live in an urban area and go to your local corporate bookstore, you will find several shelves of books devoted to describing, explaining, and providing solutions for the environmental crisis. Taking a look at these books, you will quickly notice two things. First, a large majority of them focus on climate change. Second, many of these books focus on things that individuals can do to save the environment, including changing their consumption behavior. With titles such as *Stop Global Warming: The Solution Is You!* (David 2008) and *Green Guide: The Complete Reference for Consuming Wisely* (Green Guide magazine editors 2008), some of these books focus entirely on consumer-based actions that individuals can take to solve the crisis, while others devote only a few pages or a chapter to this issue. But regardless of whether popular-press environmental books focus limited or sustained attention on individuals and consumption, many aim to convince the reader that he or she has the power as an individual consumer, rather than as a member of a social movement or social group, to solve the environmental crisis.

The argument in these books generally goes something like this: "You, the reader, probably feel overwhelmed by the immensity and severity of the environmental crisis and do not know what you can do to help solve it. However, as a consumer, you have the power to make an environmental difference. Because businesses depend on you to buy their goods and services, you can change what businesses produce and how they produce these things by committing yourself to purchasing environmentally friendly products from environmentally friendly companies. In other words, you are more powerful than the most powerful companies in the world because if they do not sell the goods and services that you want, they will go out of business. So if you and your friends go out and do your environmental duty by changing how you consume, you can help solve the environmental crisis without harming the economy or changing your lifestyles and levels of consumption."

Implied in this argument, of course, is the idea that the global environmental crisis and specific crises such as the BP oil spill are the fault of consumers and not businesses. Moreover, having bought into neoclassical economists' model of the *sovereign consumer* who acts as the final arbiter in the all-powerful market (Kirzner 1999), those who set forth this argument also clearly believe that consumers are the solution to the environmental crisis. Thus, as the quote at the beginning of this section suggests, if we want to avoid further oil-spill disasters, all we have to do is stop buying and consuming oil.

There is, of course, some logical validity to this *consumer sovereignty* argument. If every individual and organization on the planet stopped using oil, then companies would stop drilling for it, and we would have no more oil spills. However, what is logically possible and how the social world really works are two very different things. Setting aside for the moment the fact that businesses and governments also consume large volumes of goods (including oil) and therefore also affect the behavior of producers,[2] the fact that modern industrial societies are structured in such a way that it is virtually impossible for most citizens to stop using their cars (or to be able to afford electric cars), the fact that the U.S. government relies on oil and other critical natural resources for its military and geopolitical power, and the fact that government policies determine the environmental friendliness with which we dispose of goods once they become trash and enter the waste stream, this argument makes several problematic assumptions about the behavior and cognitive abilities of, and thus the relationship between, consumers and businesspeople.

For example, the consumer sovereignty argument assumes that markets provide consumers with perfect information about the goods they can potentially purchase and businesses with perfect information about the wants and desires of consumers. It further assumes that businesspeople and consumers are all rational decision makers who can use this information to optimally and efficiently achieve their goals, that consumers have a wide range of choices in the market such that there are always environmentally friendly products and services available to them that they can afford to consume, and that businesses do not successfully shape consumer demand, produce goods for which there initially is little or no demand, or attempt to conceal information about their products from consumers.

After all, if consumers do not know how goods and services are produced, what the actual environmental effects of these goods and services are, or whether there are more environmentally friendly alternatives for them to purchase, then they will have trouble effectively shaping business behavior. Similarly, if consumers are unable to cognitively process all the information necessary to rationally evaluate and accurately rank the environmental quality of various goods and services, if they use cognitive shortcuts or quick rules of thumb to make purchasing decisions, or if they sometimes let emotions shape their purchasing behavior or make spur-of-the-moment purchasing decisions, then it is unlikely that their purchases will consistently line up with their professed environmental beliefs, thereby undermining the environmental effectiveness of their purported purchasing power.

The same is true for businesses. If they do not have sufficient information to identify current consumer demand or predict future consumer demand or if they are unable to rationally process all the information available to them but instead use simplified decision making rules to choose between competing options, then it is unlikely that business behavior will be entirely shaped by consumer demand.

The reality, of course, is that businesspeople and consumers are not entirely rational and never have perfect information (Gigerenzer and Selten 2001; Kahneman 2003; Stiglitz 2002). Instead, research shows that people are only partly rational, that the decisions we make are often irrational and shaped by emotion, and that there are limits to our ability to process information, especially as the volume and complexity of the information increase (Simon 1957, 1978; Williamson 1981). This means that rather than developing rational solutions to complex problems, such as choosing between complicated options in the face of uncertainty, individuals and organizations tend to make decisions that are adequate for the task at hand (rather than making decisions that are most likely to achieve their goals) using general rules of thumb and emotional filters, or frames, that simplify information processing (Fiske and Taylor 1991; Shefrin 2002; Simon 1978; Tversky et al. 1990).

Furthermore, because the rules of thumb and emotional filters that decision makers use tend to vary according to the conditions under which they make their decisions, the context within which people find themselves can dramatically affect the choices that they make (Slovic 1995; Tversky et al. 1990). Decision makers also tend to believe that they know more than they actually do about the information on which they base their decisions, and they tend to overestimate their ability to predict their own future preferences as well as the future preferences of others (Alba and Hutchinson 2000).

These research findings clearly have important implications for the relationship between consumer demand and business behavior. In particular, they strongly suggest that consumers are unable to rationally process large volumes of complex information about the environmental quality of the goods they might potentially purchase and that what consumers want and how they translate their desires into specific purchases likely vary according to the situation in which purchasing decisions are made. It is thus unlikely that environmentally conscious consumers are able to consistently translate their desires into purchasing behavior that clearly and accurately signals these desires to producers, making it difficult for producers to respond rationally to consumer demand. These findings also suggest that even if consumers were able to send clear signals to businesses about their consumption preferences, businesses would still be unable to rationally interpret these preferences and meet consumer demand with a mix of products that give consumers what they really want.

Casting further doubt on the consumer sovereignty argument, companies and industries are often able to successfully shape consumers' desires. Data from the A. C. Nielsen Company show, for example, that the average American watches 28 hours of television per week, with the average American child watching up to 20,000 30-second commercials each year and the average 65-year-old adult viewing up to two million commercials over the course of her or his lifetime (Herr 2007). To reach these and other consumers, businesses spent almost $117 billion on advertising in the U.S. in 2009, with $440.8 billion spent on advertising around the world in 2008 (*Advertising Age* 2010).

Advertising campaigns are not always successful, of course, but often they are. Research shows, for instance, that advertising in traditional media outlets tends to increase company sales (Dertouzos and Garber 2006; Eng and Keh 2007; Rubinson 2009; Zhou et al. 2003) and that of the top 100 U.S. advertisers in 2009, those that increased their advertising spending from 2008 to 2009 were twice as likely to see their U.S. sales increase as those that spent less money on advertising in 2009 than they did the previous year (B. Johnson 2010). Research also shows that online advertising increases online *and* non-Internet sales (Lewis and Reiley 2009), that children and adults who watch food advertisements consume more food after viewing these advertisements than do other children and adults (regardless of whether they were hungry prior to seeing the advertisements) (Halford et al. 2007; Harris et al. 2009), and that as aggregate levels of advertising increase at the national level, so too do aggregate consumption levels (Jung and Seldon 1995).

Research further demonstrates that advertising most successfully promotes future sales when it relies on emotional appeals that include as little product information as possible (Binet and Field 2009; van den Putte 2009) and that advertising can shape how consumers experience the use of a product in the future as well as how they remember their prior experiences of it (Braun-LaTour et al. 2004). Moreover, indirect and vaguely stated claims are particularly effective advertising tools because they often lead consumers to make "positive inferences about . . . advertised brand[s]" that companies could not make without being accused of making false claims (McQuarrie and Phillips 2005, 7). The effect of indirect and vaguely stated claims is particularly powerful when the claims are made visually rather than verbally, leading McQuarrie and Phillips (2005) to argue that visual advertising that uses such claims can easily mislead consumers into believing a product has qualities it does not have. For example, a picture of a pristine natural environment on a bottle of cleaning detergent may lead consumers to believe that the detergent is good for the environment even if it harms the environment.

Vague and misleading environmental claims can be found on a wide variety of product packages and labels. For instance, in a recent study of 2,219 green products found in leading "big box" retail stores in the U.S. and Canada, TerraChoice (2009) found that more than 98% of the 4,996 pro-environmental claims made by these products were vague or misleading. TerraChoice refers to these claims, which fall into seven broad categories, as "greenwashing sins." One of these sins, the *sin of worshiping false labels*, refers to "an effort by some marketers to exploit consumers' demand for third-party certification [by providing consumers with] fake labels or claims of third-party endorsement" (TerraChoice 2009, 4). Examples of this sin include "a brand of aluminum foil with certification-like images that bear the name of the company's own in-house environmental program without further explanation . . . [and] a variety of products . . . [that] use certification-like images with green jargon [such as] 'eco-safe,' 'eco-secure,' and 'eco-preferred'" (TerraChoice 2009, 5).

Another example of potentially misleading third-party certification labels can be found on the packaging of many fair trade organic coffees, the product of an industry that claims to "empower poor coffee growers, provide [these] coffee growers with a living wage, reduce rural poverty, support environmentally sustainable farming practices, and improve food security" (Jaffee 2007, 3). However, according to Daniel Jaffee (2007), who spent nearly two years interviewing local coffee growers and community leaders in Yagavila and Teotlasco, Mexico, the benefits that poor, developing nation farmers receive for growing organic fair trade coffee are somewhat mixed. On the one hand, organic fair trade coffee growers in Yagavila and Teotlasco have higher incomes and greater food security than do other coffee growers in these communities. They also use more environmentally beneficial farming practices than do other local coffee growers. On the other hand, supplying organic coffee to the fair trade industry does not pull these farmers out of poverty, it requires a tremendous amount of extra work due to the strict requirements of organic certification, and the compensation that farmers receive does not always cover the costs of production. The prices the industry pays these farmers have also remained nearly constant since the inception of the industry and are lower than the minimum price consumers are led to believe the farmers receive. Thus, rather than benefiting in proportion to their effort, it appears that these farmers are being asked to unfairly subsidize the environmental concerns of northern consumers. Not surprisingly, many coffee growers in these communities refuse to grow organic fair trade coffee.

Further complicating matters, Jaffee notes that TransFair USA, the only third-party certifier of fair trade products in the U.S. as of early 2010, gives its fair-trade organic coffee certification endorsement to antiunion companies such as Starbucks, to companies whose fair trade coffee sales represent only a small percentage of their overall coffee sales, and to plantations that use wage labor, all of which seem to contradict the fair trade movement's claims regarding the importance of peasant and worker empowerment. The standards that TransFair uses to determine whether a company should receive its fair trade label and the agreements it signs with companies that want its fair trade endorsement also appear to vary according to which company wants the endorsement, with larger companies apparently receiving better treatment (less strict terms) than do smaller companies.

Jaffee's research thus suggests that consumers who purchase organic fair trade coffee are probably not buying what they think they are buying and that they, like consumers of other misleadingly labeled products, are not doing a very good job of shaping business practices in a manner that meets their consumption goals. Jaffee's research, the Terra-Choice study, and McQuarrie and Phillips's findings regarding vaguely stated advertising claims further indicate that businesses often meet pro-environmental demand by pretending to give consumers what they want or by leading consumers to believe that their purchases are better for the environment than they really are. This, of course, contradicts the consumer sovereignty argument, which, as previously noted, assumes not only that decisions makers are fully rational actors who have access to perfect information but also that businesses do not successfully shape consumer demand by misleading consumers about their products.

The final consumer sovereignty assumption I want to examine is the assumption that companies do not produce goods that eventually sell well but for which there initially is little demand, no demand, or a long-term decline in demand. This assumption is important because if companies do produce such goods, it suggests that their production strategies are not based solely on consumer demand, that they produce goods for which they probably recognize they will have to create demand,[3] and that they are able to successfully create demand and thus shape consumers' desires.

Contradicting this assumption, and thus the consumer sovereignty argument, a fair number of studies show that companies often produce goods and services for which there initially is no demand, little demand, or a long-term decline in demand. For example, the Sony Walkman, which became very popular in the 1980s, was created by the tape recorder division of Sony because the division had lost control of one of its key products (radio-cassette recorders) to the radio division and had to develop a new product to maintain its profitability. Thus, the Sony Walkman was not invented in response to consumer demand but instead because the tape recorder division was worried about its survival and because the division and corporate leadership thought (correctly but without any proof) that young consumers would like it (du Gay et al. 1997).

In a somewhat different vein, the men's clothing industry in the U.S. overcame declining sales in the 1960s by convincing young men that clothing was a form of self-expression that could be used to signal the wearer's creativity and lack of conformity (Frank 1997). Similarly, the coffee industry overcame steep declines in its sales in the 1960s, 1970s, and 1980s by convincing specific segments of the public that consuming specialty coffees fit their lifestyles (Roseberry 1996). Although it is true that the clothing and coffee industries were both responding to and taking advantage of consumers' desires to express their unique identities, in neither case were they responding to consumer demand for specific products. Instead, they successfully created new demand for products that consumers were no longer as interested in buying as they had been in the past.

Finally, in a highly compelling case study, Tim Bartley (2007) demonstrates that the forest certification industry, which provides certification for wood grown in environmentally and socially sustainable ways, was created not in response to consumer demand but rather by a handful of large U.S. foundations, including the MacArthur Foundation, the Ford Foundation, and the Pew Charitable Trusts, that wanted to create a less radical alternative to the tropical timber boycotts proposed by many anti-logging activists. These foundations soon realized, however, that creating a forest certification industry would be pointless if consumers did not know about or want their product. They therefore created demand for certified forest products by (a) helping organize a buyers group that worked with companies such as Home Depot, Ikea, the Gap, and Starbucks and (b) providing this and another buyers group with more than $4 million in funding between 1996 and 2001. Bartley thus concludes that "consumer demand [for certified forest products] had to be *mobilized and organized*, and that foundations were at the forefront of this process" (2007, 244; emphasis in the original).

Bartley's findings clearly contradict the consumer sovereignty argument. Further undermining this argument, not all forest certification programs actually protect the environment. The Natural Resources Defense Council noted in 2010, for example, that worldwide there are at least nine forest certification programs, eight of which "are backed by timber interests and set weak standards for forest management that allow destructive and business-as-usual forestry practices" (NRDC 2011).

It thus appears that the consumer sovereignty argument is poorly supported by the evidence. This is the case, I argue, because corporations are much more powerful than consumers due to their vast financial and political resources, which, when combined with consumers' lack of time, limited information gathering resources, and cognitive weaknesses, allow corporations to shape consumers' desires and control what consumers know about their products and production processes.

As many researchers have noted, the ability of a group or organization to shape what others want and what they know about a subject is an important form of power (variously described as knowledge, information, or value power), the possession of which increases the likelihood that the group or organization will achieve its goals either in the face of resistance from others or by forestalling resistance through the creation of organized ignorance (Bonds 2011; Boyce 2002; Freudenburg and Alario 2007; Gaventa 1980). And since what many corporations want to do is sell products that are environmentally harmful or that are produced in environmentally and socially harmful ways, it should come as no surprise that they often use their knowledge, information, and value power to achieve this goal. What is surprising is that proponents of the consumer sovereignty argument neglect these facets of power and inequality, arguing instead that individual consumers have more power than do corporations that sell tens to hundreds of billions of dollars of goods and services each year. This is an erroneous argument that leads these proponents to misdiagnose the fundamental causes of the environmental crisis and set forth unworkable solutions to the crisis.

But if individuals are unable to translate their desires into effective consumer demand and businesses are unable or unwilling to produce the goods and services that consumers desire, is there some way to compel them both to behave in more environmentally friendly ways, even if neither is interested in doing so? Environmental economists argue that they have found a way to do just this.

Environmental Economics

Environmental economists study a broad array of topics, the most important of which is the role that markets and prices play in shaping environmental outcomes.[4] Specifically, they argue that properly functioning markets provide us with the most efficient means of allocating scarce resources but that capitalist markets do not function well because they do not take into account all the environmental and social costs of production and consumption.[5] As a result, these costs are not included in the prices of the goods and services that businesses sell. This is problematic, they argue, because it makes environmentally and socially harmful goods and services less expensive than they otherwise would be. It also makes them less expensive than products that are produced in socially and environmentally sustainable ways, leading businesses to produce, and businesses, governments, and consumers to purchase, products that are more, rather than less, harmful to society and the environment. Not incorporating social and environmental costs into prices is also problematic, they argue, because in many instances, the people, companies, and governments that end up paying these costs (either financially or otherwise) are not the ones responsible for creating them or the ones that benefit most from the activities and products that produced them.

For example, when companies manufacture products such as cars and petrochemicals, neither they nor consumers normally have to pay directly, as part of the products' prices or the business' costs of production, for the health care of people who are not employed by the company but whose health is negatively affected by the production and consumption of these goods (nor do they pay directly for most of the work-related health care costs of company employees who do not receive employer-paid health insurance).[6] Similarly, neither companies nor consumers pay directly, again in terms of prices and production costs, for most of the social and environmental harm caused by resource extraction activities. For instance, oil companies and consumers do not pay directly, as producers or consumers, for the costs of global warming or the costs of wars that are fought to maintain oil companies' access to oil supplies.

Of course, many of these costs are paid by someone, including taxpayers, insurance company policyholders, people who are poisoned by petrochemicals and car exhaust, the hundreds of thousands of people around the world who have been forcibly removed from their homes or have lost their livelihoods to make way for resource extraction activities, soldiers who fight in oil wars, and people who live in areas where oil wars and other natural resource conflicts occur. But the important point for understanding environmental economics is that these costs are not directly incorporated into the prices that businesses, governments, and consumers pay for the products they consume, which means that the costs of these products are kept artificially low and the demand for them artificially high.

Not surprisingly, then, environmental economists argue that to solve the world's many social and environmental crises, we must incorporate the social and environmental costs of producing, transporting, and consuming goods and services into the prices of these goods and services. This will ensure on the one hand that socially and environmentally harmful products are more expensive than socially and environmentally sustainable products and on the other hand that those who produce, purchase, or otherwise benefit from socially and environmentally harmful products will have to pay (at least financially) for the role they play in harming other people and the environment. More importantly, increasing the prices of environmentally and socially harmful goods will result in decreased demand for these goods and increased demand for sustainably produced goods, leading eventually to a situation in which goods and services are produced and consumed in much more environmentally and socially sustainable ways.

Moreover, as more and more companies begin to produce socially and environmentally sustainable products and as the volume of production of these products increases, their price should decline due to the competitive pressures of the marketplace and savings realized through economies of scale. Environmental economists thus predict that if their policy prescriptions are adopted, the prices that consumers, businesses, and governments pay for goods and services will increase and then decrease, though it is not entirely clear that prices will ever drop to their pre-policy adoption level.

Environmental economists have proposed a variety of policies that governments can adopt to incorporate social and environmental costs into prices. They note, for example, that governments can place levies on products that cause environmental damage, provide subsidies to companies that produce environmentally sustainable products (which lowers the prices of these products), or tax companies on the basis of the amount and types of pollution they emit or the volume of natural resources they consume. Governments can also create cap-and-trade systems in which pollution caps are combined with permits to pollute that can be traded between companies and/or nations (I provide a detailed description of cap-and-trade later in the chapter).

Environmental economists further argue that their market-based approach is superior to government regulation because government regulation is cumbersome, unwieldy, and expensive due to the need for high levels of government oversight and the slow-moving and inefficient nature of government bureaucracy. Government regulation is also problematic, they argue, because (a) government regulators are often strongly influenced by industry, (b) businesses often provide government agencies with inaccurate information, and (c) government agencies often lack the expertise, resources, and technical and proprietary information they need to effectively oversee industry. Environmental regulations are also easily evaded, they contend, and provide companies with no incentive to reduce pollution below mandatory levels.

Another reason why environmental economists prefer market-based solutions over government regulation is that market-based solutions supposedly produce economically efficient outcomes, outcomes in which resources are allocated among productive end uses in such a way that (a) goods are produced at their lowest cost and (b) reallocating the distribution of goods and services to make one person better off cannot be done without harming someone else. Since economists argue that efficiency can only be achieved when prices are set in competitive markets by the forces of supply and demand, it thus follows that for economists, market-based solutions to environmental problems are necessarily more efficient and effective than are government regulations, which, they argue, distort prices by impinging on the forces of supply and demand in the market.

There is, I believe, much to admire about environmental economics, in particular its emphasis on internalizing the many social and environmental costs that are currently ignored in our system of price determination and its emphasis on using a single type of policy mechanism to shape both business and consumer behavior.[7] However, despite its strengths, mainstream environmental economics is beset by a number of problems, some more serious than others, that suggest not only that its policy instruments will be less effective than advertised but also that it does not address the fundamental causes of the world's major social and environmental problems or explain why businesses currently fail to incorporate social and environmental costs into the prices of their goods and services.

One problem with environmental economics is that if we follow its policy prescriptions, the increased prices that we pay for goods and services in the short run (and possibly the long run) will disproportionately burden lower- and middle-income families. Measures can be taken to mitigate this disproportionate burden, but the factors that determine whether policy makers incorporate such measures into their policy instruments fall outside the purview of environmental economics and, thus, are not explained by the theory or shaped by its policy instruments.

A second problem with environmental economics is that it is not entirely clear that its policy prescriptions are nearly as efficient as environmental economists claim they are or that they avoid the problems of government oversight and inefficiency that supposedly hamper government regulation. To the contrary, environmental economists' policy instruments almost universally require government intervention in the market to address the problem that markets do not "naturally" incorporate social and environmental costs into prices. For example, taxes are a form of government intervention that according to economic theory reduce market efficiency by impinging on the forces of supply and demand. Tax payments must also be monitored and enforced by a supposedly inefficient government bureaucracy using data that is either collected by the government or provided to the government by private actors. Providing subsidies to industry for meeting specific environmental performance goals is also a form of government intervention into the market that both impinges on supply and demand and requires government officials to efficiently monitor business performance using data collected by or provided to government agencies.

Moreover, because it is virtually impossible to accurately determine the dollar value of the social and environmental costs associated with producing and consuming goods and services, the best we can do is approximate these costs. We obviously do not need to put an accurate dollar value on these costs to shape consumer and business behavior, but accurately determining these costs does appear to be necessary if economic policy instruments are to produce the kinds of efficient outcomes that economists claim that they will. After all, environmental economics is based in large part on the argument that capitalist markets do not allocate efficiently when prices are inaccurate. Moreover, it will be government bureaucrats or politicians, and not market actors and the forces of supply and demand, that determine the dollar value of these costs or, in the case of cap-and-trade systems, the pollution caps that shape these costs.

It is difficult to understand, therefore, why environmental economists' policy solutions are necessarily more efficient or effective than government regulation or to understand how the problems supposedly associated with government regulation will be solved by developing policy instruments that ask governments to tax, subsidize, levy charges, and develop, monitor, and enforce complicated emissions trading schemes.

Another fundamental problem with environmental economics is that it largely ignores power and inequality. Public policies are not created in a vacuum or with an eye toward maximizing the welfare of as many people as possible or those who are most in need. Instead, government policies are the product of political struggle regarding which issues make it onto the political agenda, which policies to implement, how these policies will be monitored and enforced, and who these policies will benefit the most and hurt the least (Gaventa 1980; Domhoff 1990, 2002). And because public policies are the outcome of political struggle, they tend to be shaped most strongly by those actors who have the most economic and political power (Burris 1992; Domhoff 1990, 2002; Dreiling 2001). As a result, whether environmental economists' policy proposals are adopted, the specific form that these policies take, how effective they will be in achieving progressive social and environmental goals, and who they will most benefit and harm are determined in the realm of politics, which is shaped in fundamental ways by economic and political inequality, subjects about which environmental economics has virtually nothing to say.

To better illustrate this argument, I devote the remainder of this section to a discussion of cap-and-trade systems for greenhouse gas emissions. Many solutions have been proposed by environmental economists to solve the climate change crisis, including taxing carbon emissions and fossil fuel use, subsidizing the use of alternative energy technologies, and creating cap-and-trade emissions systems (Nader and Heaps 2008; Nordhaus 2009). Of these various solutions, cap-and-trade systems have received the most political support in both the United States and various global fora. For instance, the European Union adopted cap-and-trade as one of its key mechanisms for solving the climate change crisis (Ellerman and Joskow 2008), as did the U.S. House of Representative's American Clean Energy and Security Act of 2009 (otherwise known as the Waxman-Markey bill), which was supported by the Obama administration and approved by the House of Representatives but not the Senate (Wessel 2009).

One possible reason for why cap-and-trade systems have been the preferred policy option for solving the climate change crisis may be that such systems are superior to other approaches in terms of their enforceability, cost, efficiency, and effectiveness in reducing greenhouse gas emissions. However, as described in the following pages, this does not appear to be the case. Moreover, the cap-and-trade systems adopted in the European Union and the Waxman-Markey bill appear to be particularly ineffective and inefficient forms of cap-and-trade.

Cap-and-trade systems, in their simplest form, impose a regulatory limit on the quantity of specific pollutants that can be emitted within a country or other administrative unit and then provide polluters with permits to pollute, with each permit equivalent to a specified quantity of the regulated pollutant (in greenhouse gas emissions systems each permit generally allows polluters to emit one ton of carbon dioxide or its equivalent). Polluters that are able to affordably reduce their pollution below their permitted levels can then sell their excess permits to polluters that find it too expensive to do so, thereby allowing emissions to be reduced in the least expensive way possible.

Of course, cap-and-trade systems are never as simple as this. For example, in an article from the early 1990s, Adam Rose and Thomas Tietenberg (1993) list a whole range of decisions that must be made, including those having to do with government oversight activities, to make cap-and-trade systems operational. These decisions include, but are not limited to, the following: Are permits to be given at the extraction, production, consumption, or emission stages of a product's life cycle? Are they given for a single year or multiple years? How are permits to be allocated among countries and within countries? When initially created, are permits given away or auctioned off? If companies take actions such as planting trees that absorb greenhouse gases, will such actions be considered "offsets" that increase the quantity of greenhouse gases they can emit? If so, who determines whether offsets have really occurred and what rules are used to make these determinations? Is the trading of permits conducted through a centralized clearinghouse, and if so, who runs this clearinghouse? Can permits be traded on spot and derivatives markets? What government agencies will monitor greenhouse gas emissions to ensure that those who do and do not hold permits follow the rules? How will emissions be monitored? What government agencies will enforce compliance? How will governments monitor permit trades and leases so as to create accurate emissions inventories? What government agency will certify that permits are valid before they are traded? What international monitoring agency will be created to ensure that nations are following the rules? What individuals, organizations, and government actors will have standing to file complaints for noncompliance by businesses and nations? What agency will hear these complaints?

This is a fairly lengthy, albeit incomplete, list of decisions and government oversight activities that must occur to make cap-and-trade systems operational, and it suggests quite strongly that cap-and-trade systems require at least as much government intervention, government oversight, and information flow between businesses and government as do most, if not all, government regulatory systems. It certainly requires much greater government capabilities and larger volumes of information flow than does Nader and Heaps's (2008) proposal to solve the climate crisis by "apply[ing] a carbon tax ... at a relatively small number of major carbon bottlenecks,[8] which cover the lion's share of [greenhouse gases] ... [and] can be monitored by satellite and checked against the annual surveillance of fiscal and economic policies already carried out by IMF staff."

It thus appears that the most favored environmental economics policy option is not particularly efficient, simple, straightforward, or free of government intervention, thus undercutting most of the advertised benefits of this supposedly market-based solution. Moreover, the fact that cap-and-trade has received much more support from policy elites than have fossil fuel or emissions taxes despite the relative simplicity of Nader and Heaps's policy proposal and the fact that fossil fuel and emissions taxes are likely to be more efficient and effective than cap-and-trade (Nader and Heaps 2008; Nordhaus 2009; Shapiro 2009) suggests that factors ignored by environmental economics play an important role in determining which environmental policies are seriously debated and adopted.

In the case of cap-and-trade, the most prominent factors shaping the U.S. debate appear to be corporate power (*Economist* 2009) and politicians' fears of voters' and campaign funders' reactions to higher taxes (Carlson 2009; *Economist* 2009). For example, according to the *Economist* (2009), the 2009 Waxman-Markey bill was remarkably similar to a bill proposed in 2007 by USCAP, an industry group whose members would have benefited greatly from a cap-and-trade system. This, the *Economist* argues, suggests that "business . . . was largely responsible for Waxman-Markey's inception" (2009). The *Economist* goes on to argue that intense opposition to the Waxman-Markey bill by oil companies and energy-intensive manufacturers played a strong role in generating political opposition to the bill in the U.S. House and Senate. Thus, according to the *Economist*, the actors that had the greatest positive and negative influence on the Waxman-Markey bill were business groups that had strong financial interests in whether the bill became law.

The lengthy list of decisions and government oversight activities just presented also demonstrates that developing a cap-and-trade system is a fairly complicated process in which many decisions have to be made that will likely affect not only the efficiency and effectiveness of the system but also who benefits most and least from it. For instance, both the European Union (EU) cap-and-trade system and the system proposed by Waxman and Markey allow the vast majority of pollution permits to be given away rather than auctioned off. However, as White House budget director Peter Orszag noted before the Waxman-Markey bill was voted on, if permits are given away for free, "all of the evidence suggests that . . . corporate profits [will] increase by approximately the value of the permit[s]" (Wessel 2009). Moreover, in EU nations with deregulated electricity markets, electricity companies that received free permits were, as of 2008, apparently able to pass much of the market price of these freely received permits on as a cost to consumers, who then had to pay for what the companies had received for free (Ellerman and Joskow 2008). Thus, intentionally or not, the decision to give permits away for free is also a decision about how the costs and benefits of cap-and-trade are to be distributed across social groups.

The EU emissions system and the Waxman-Markey bill also included provisions allowing permits to be traded on commodity markets as asset-based financial instruments (Monast et al. 2009), with some industry experts estimating that the value of these markets would have been $3 trillion by 2020 if the U.S. had adopted a cap-and-trade system in 2009 (Gillibrand 2009). Not surprisingly, Wall Street bankers and investors saw this as a potentially "lucrative source of fees, commissions, and speculative gains" and thus were among the strongest supporters of cap-and-trade (Shapiro 2009).

How the right to pollute is allocated across nations also has distributional consequences. For example, under the Kyoto Protocol, wealthy regions and nations such as Europe, Japan, Canada, and Australia were given the right, depending on the region and nation, to emit between 92% and 108% of their 1990 greenhouse gas emissions until 2012 (Lohmann 2001), thereby providing them with the right to continue polluting and, if they adopted an emissions trading system, to distribute to their nations' businesses potentially valuable pollution permits. It is quite understandable that wealthy nations did not want to enact drastic emissions cuts during the first few years of the Kyoto Protocol. However, the Protocol could have given each signatory nation a set of pollution rights proportional to that nation's population size such that wealthy nations and corporations would have initially been able to pollute at relatively high levels but would have had to pay developing nations for much of this right (by buying developing nations' permits) rather than receiving it for free.

Finally, the permit systems set up by the EU and included in the Waxman-Markey bill both allowed businesses to invest in pollution offsets. Offsets work as follows: if a company invests in the development of an overseas wind farm, reforestation project, hydropower plant, environmentally clean industrial facility that replaces a dirtier facility, or some other similar project that reduces the volume of greenhouse gases in the atmosphere, it can use the carbon credits it receives from these investments to increase the amount of greenhouse gases it emits (in other words, the company's extra emissions are offset by the reduction in greenhouse gases resulting from the "offset" investment). The idea behind this is that in many instances, offsets allow companies to reduce their climate change impact more inexpensively than they otherwise could (Carlson 2009), thereby reducing the overall cost of a nation's cap-and-trade system.

Offsets are quite problematic, however. First, it is not clear that scientists know enough about carbon flows to know how much carbon is captured or eliminated in many offset projects. Second, offsets can be awarded for reducing environmentally harmful activities in a specific location, but that does not mean that these harmful activities will not simply move elsewhere. For instance, reducing or eliminating forest clearcutting in one region of a country does not mean that loggers will not commence clearcutting in another region of the country or in another part of the world (Lohmann 2001). Third, as of 2007, firms that certified offset credits for the European Union operated with few detailed standards and guidelines, faced few sanctions for poor performance, and were spending less and less time certifying individual projects due to high levels of interfirm competition that had decreased the prices these firms received to conduct offset certifications (Schneider 2007). Fourth, offsets are only supposed to be awarded for projects that would not otherwise be undertaken so as to ensure that credits are given only for investments that contribute to a reduction in atmospheric greenhouse gases. Known as *additionality*, this condition often goes unmet even when emissions credits are awarded (Victor 2009).

As a result of these and other problems, several studies have shown that offsets often do little or nothing to reduce greenhouse gas levels in the atmosphere (Victor 2009). For example, Lambert Schneider (2007) found that in the European Union in 2007, roughly 20% of offset credits lacked additionality; and Michael Wara, an environmental lawyer at Stanford University who specializes in climate change policy, argues that determining whether additionality has occurred is virtually impossible (*Economist* 2009). This is particularly troubling because as of 2008, more than two-thirds of the European Union's stated emissions reductions had come from offsets and less than one-third from actual emissions reductions (Nordhaus 2009).

Since offsets play such a large role in Europe's efforts to reduce greenhouse gas emissions and since they would have played a much bigger role in U.S. efforts had the Waxman-Markey bill become law (Carlson 2009; *Economist* 2009; EPA 2009), these studies thus suggest that neither of these emissions systems was designed so as to overcome climate change. This, of course, begs the question of why these systems were designed the way they were. The likely reason is that without offsets the economic costs of these plans would have increased dramatically (Carlson 2009; EPA 2009), making them much less palatable to politicians seeking both reelection and the support of business. In other words, it is likely that political calculations and the power of business trumped environmental considerations in this instance, as they likely did when choices were made about how to allocate permits and pollution rights within and across nations, whether to give away or sell permits, whether to allow permits to be traded as asset-based financial instruments, and whether to choose cap-and-trade as the most viable approach for addressing global climate change.

Conclusion

In the preceding section, I demonstrated that environmental economists' policy proposals are not necessarily any more efficient, effective, or devoid of government intervention than are government regulations and that in some cases they are likely to be less efficient and effective. This is not to say that environmental economic policy instruments are never more effective than government regulation but instead that the advantages and disadvantages of market-based policies and government regulation need to be determined based on an evaluation of specific policies rather than on a general theoretical argument that holds that government intervention is bad and that market-based policies are necessarily efficient and free of government intervention.

The evidence I presented in the preceding section also strongly suggests that inequality and power play an important role in determining which policies governments adopt, how effective these policies will be in achieving progressive social and environmental goals, who these policies will most benefit and harm, and how strongly these policies will shape business behavior. I have not proven this general point yet, but it certainly appears that the decisions that were made when crafting the European Union and Waxman-Markey emissions schemes tended to favor powerful economic interests, often to the detriment of poor nations, the environment, and average citizens; and as G. William Domhoff (2002) demonstrates, routinely benefiting from government policy is an important indicator of political power. Furthermore, the high level of correspondence that existed between the Waxman-Markey bill and the bill proposed by the industry group USCAP and the role that oil companies and energy-intensive manufacturers played in creating opposition to the Waxman-Markey bill in the U.S. House and Senate both suggest that corporate interests disproportionately influence U.S. policy making.

If inequality, power, and corporate interests do play important roles in shaping economic and political outcomes, and much of the evidence presented later in the book demonstrates that they do, then environmental economics does not address the fundamental causes of the world's major social and environmental crises because the factors that determine whether policy makers adopt its policy proposals (or other pro-environmental policy proposals) and whether businesses incorporate social and environmental costs into prices fall outside its purview and, thus, are not explained by the theory or shaped by its policy instruments.

As I argued earlier in the chapter, the consumer sovereignty argument also fails to take inequality and power into account, which results in its proponents assuming that individual consumers are able to shape business behavior, an erroneous assumption that leads these proponents to misdiagnose the fundamental causes of the environmental crisis and set forth unworkable solutions to the crisis.

Unfortunately, these two prominent and influential explanations of the environmental crisis are not unique among mainstream environmental explanations in ignoring inequality, power, and the social structures that maintain inequality and power. Thus, most mainstream environmental thought and action are motivated by a highly inaccurate understanding of how the social world operates, leading to an environmentalism that is in most cases fairly ineffective. It is no wonder that one of this nation's leading environmentalists, James Gustave Speth, in his 2005 book *Red Sky at Morning*, argues so forcefully that despite the environmental movement's achievements, future generations are going to wonder why the movement did so little to save the environment.

A better approach to explaining the global environmental crisis, the macro-structural environmental sociology approach, addresses this inattention to inequality and power by focusing on the social structural causes of environmental problems. In doing this, it addresses some of the major limitations of the consumer sovereignty and environmental economics arguments, as well as the shortcomings of other explanations of the environmental crisis that ignore or downplay inequality and power. Macro-structural environmental sociology is not without its own limitations, however. Thus, in the next chapter, I describe the basic tenets of macro-structural environmental sociology and argue that its practitioners generally pay insufficient *theoretical* attention to undemocratic institutions and elite-controlled organizational networks. I then set forth my own macro-structural theoretical model, the inequality, democracy, and the environment (IDE) model, which holds that organizational, institutional, and network-based inequality plays a fundamentally important role in harming individuals, societies, and the environment by giving elites the power to develop and control organizational, institutional, and network-based mechanisms that they use to achieve their goals in the face of resistance from others. It is important to note, however, that my goal in setting forth this model is not simply to fill several critical gaps in the macro-structural environmental sociology literature. More importantly, I hope to increase researchers' and environmentalists' understanding of the social structural roots of the environmental crisis and to help mainstream environmentalists and activists more easily identify the organizations, institutions, and networks that need to be reformed or abolished if we are to solve the myriad social and environmental crises facing us.

2

Inequality, Democracy, and Macro-Structural Environmental Sociology

I have three goals in this chapter: to describe the strengths and weaknesses of macro-structural environmental sociology, to set forth my own theoretical explanation of the environmental crisis, and to situate my theoretical explanation within the broader macro-structural environmental sociology tradition.[1] However, to best understand this tradition, it first helps to know what sociologists mean when they use the term *social structure*.

Social structure refers to the ways in which societies are organized and how the particular organization of any society shapes (a) the attitudes, values, beliefs, and behavior of individuals, groups, and organizations within the society, (b) the opportunities and choices available to these individuals, groups, and organizations, and (c) social outcomes such as the levels of inequality and crime in the society, the prevalence of divorce, disease, and teenage childbirth in the society, and the levels and types of environmental degradation produced by the society. Social structures—which include but are not limited to the voting systems that democratic nations use, the laws and regulations governing market activity, the international treaty system, and the informal rules that guide human interaction—exist at the local, national, and global levels, are found in all realms of social life, and organize all human activity. As a result, they play a critically important role in shaping all social outcomes, all social relationships, and all individual, group, and organizational behavior.

Not surprisingly, macro-structural environmental sociologists argue that local, national, and global structural arrangements are the most important set of factors shaping the global environmental crisis. This is so, they contend, because these arrangements reward environmentally destructive behavior, limit the pro-environmental behavioral choices available to individuals, communities, and organizations, and produce attitudes, values, and beliefs that are antithetical to environmental sustainability.

The two most prominent theoretical perspectives within macro-structural environmental sociology are the treadmill of production theory and the world systems theory (WST) approach to understanding ecological issues. Thus, I spend the remainder of this section summarizing and critiquing these two theories.

According to treadmill theory, the competitive pressures of capitalism force businesses to continually seek out new ways to generate profits and accumulate capital, which they do by expanding their operations and investing in labor-saving technologies that reduce labor costs but increase energy and chemical use (Schnaiberg 1980; Gould et al. 2004, 2008). Local, state, and national governments support, subsidize, and legitimate these activities in order to maintain tax revenues and their own legitimacy, the theory holds,[2] and workers support economic expansion because economic expansion promises to replace jobs lost to technological innovation (Obach 2004). However, economic expansion and technological innovation are environmentally destructive because they result in increased pollution, increased resource use and extraction, and reliance on ever more toxic and hazardous substances (York 2004). Economic expansion and technological innovation are also *socially* destructive because the shift to labor-saving technology that accompanies economic expansion results in an ever smaller share of society's economic surplus going to workers, thereby harming the working and middle classes, increasing class inequality, and solidifying the power of capitalists. The irony that treadmill theorists thus highlight is that the only groups in capitalist societies that have an interest in confronting and challenging the capitalist treadmill are the working and middle classes (and the unemployed), who, by virtue of their continually increasing dependence on capital, end up regularly calling for an expansion, rather than an overturning, of the capitalist treadmill.

Treadmill theory thus focuses researchers' attention on structural arrangements in the economy that make workers and governments dependent on capitalists and that put pressure on companies to adopt labor-saving technology and expand production. In highlighting worker and government dependence on big business, as well as the inability of workers to prevent job losses arising from technological innovation, the theory also emphasizes the important role that inequality and power play in degrading the environment. Indeed, treadmill theorists view rising inequality as both a key consequence of and a critical driving force behind the ever expanding capitalist treadmill.

The world systems theory approach to ecological issues, which I call ecologically oriented WST, or environmental WST, highlights a different set of structural factors than does treadmill theory. Proponents of world systems theory hold that capitalism is best understood as a dynamic world system composed of core, periphery, and semi-periphery nations. Core nations are the world's wealthiest, most industrialized, and most powerful nations; periphery nations are the world's poorest, least industrialized, and weakest nations; and semi-periphery nations fall in between the two. According to the theory, core nations' wealth and power, periphery nations' poverty and underdevelopment, and the inability of most semi-periphery nations to attain core nation status result from core nations' current and historical exploitation of periphery and semi-periphery nations' labor power and natural resource wealth. This exploitation, along with the structure of economic and trade relations that characterize the capitalist world system at any point in time, shapes the behavior of core, periphery, and semi-periphery nations as well as the opportunities and constraints they experience. Whether a nation belongs to the core, periphery, or semi-periphery also affects the types of goods that nation produces, the technology it uses to produce these goods, its trade relations with other countries, and whether foreign investment in its economy and international trade affect it positively or negatively.

Although not originally crafted to address environmental issues, world systems theory has been profitably employed to do just that. For example, J. Timmons Roberts and Bradley Parks (2007) demonstrate that world system structural constraints shape nations' susceptibility to climate-based disasters as well as the negotiating positions that national leaders take vis-à-vis climate change; and a growing body of research demonstrates that international trade (Jorgenson and Burns 2007; Shandra et al. 2008) and foreign investment (Jorgenson 2007, 2009; Kentor and Grimes 2006) are associated with higher levels of greenhouse gas emissions, deforestation, and air and water pollution in developing nations than in other nations due to developing nations' structural position in the world economy.

A new and important variant of environmental WST, ecological un-equal exchange theory, has also received much attention in recent years. Ecological unequal exchange theory holds that core nations are not only able to take advantage of the natural resource wealth of periphery nations through mining and other environmentally degrading resource extraction activities, they are also able to export many environmentally degrading manufacturing and agricultural activities to the periphery, all of which results in increased environmental degradation in the periphery and reduced environmental degradation in the core (Jorgenson and Clark 2009). This research thus shows that structural position in the world system hierarchy plays an important role in shaping nations' ecological exchange relations (the flow of ecological wealth and degradation across national borders), with important consequences for nations' ability to minimize environmental degradation within their own borders and take economic advantage of their own and global natural resources (J. Rice 2009; Jorgenson 2009).

The treadmill model, environmental WST, and ecological unequal exchange theory provide social scientists and environmentalists with many critical insights into both the structural relationship that exists between humans and the environment and the role that capital accumulation, the pursuit of profits, inequality, and power play in promoting environmental degradation. Nevertheless, like all theories, these theories privilege certain features of social life and certain aspects of the social structure over others. For example, although treadmill theorists identify three structural mechanisms that are both central to their theoretical model and used by capitalists to accumulate capital—expanding productive capacity, using labor-saving technology, and taking advantage of the structural correspondence that exists between capitalists interests and those of labor and the state—these are not the only mechanisms that capitalists and states use to foster capital accumulation. Other mechanisms include international trade agreements such as the North American Free Trade Agreement (NAFTA), international trade and finance institutions such as the World Trade Organization (WTO), the World Bank, and the International Monetary Fund (IMF), and corporate-controlled commodity chain networks.[3] As we shall see later in the book, these treaties, organizations, and networks, and others like them, promote capital accumulation, human suffering, and environmental degradation by, among other things, enforcing unequal terms of trade, ensuring corporate access to natural resources, increasing corporate control over production and trade networks, and maintaining global inequality (Bello et al. 1999; Carolan 2008; Wallach and Woodall 2004). These treaties, organizations, and networks are also part of a larger institutional structure, or *structure of accumulation*, that is not emphasized by the treadmill model but that provides corporations and investors with the stability, predictability, and opportunities they need to accumulate capital (Fligstein 2001; Gordon et al. 1982).

Another limitation of the treadmill model is that it focuses relatively little *theoretical* attention on the role that undemocratic and elite-controlled organizations, institutions, and networks play in promoting capital accumulation and environmental degradation. Focusing theoretical attention on undemocratic institutions and elite-controlled organizational networks is critically important, however, because institutions, organizations, and networks are among the most important building blocks of social structure, in large part because they are the arena in which and conduit through which power is exerted and economically, politically, and environmentally important decisions are made and enforced. As a result, theoretical arguments that pay little or no attention to organizations, institutions, and networks tend to have an abstract quality to them in which social outcomes are the product of purely impersonal forces rather than of humans acting intentionally to achieve their goals, albeit within organizational, institutional, and network-based social structures and economic, political, and ideological systems that fundamentally shape their interests, goals, and capacities to achieve their goals.

The treadmill model also relies on a relatively abstract theory of capital-state relations (the relations that exist between capitalists, elected officials, and government bureaucrats) that essentially holds that the government's dependence on capital is so strong that capitalists get what they need from the government without having to directly involve themselves in policy making. Research demonstrates, however, that to achieve their goals, capitalists do have to directly involve themselves in the policy making process (Domhoff 1990, 2002; also see chapters 3–6). As a result, theorizing the ways in which capitalists do this is of critical importance if we are to fully understand the macro-structural roots of the current environmental crisis, the role that economic and political elites play in harming the environment, and the mechanisms that elites use to achieve their goals.

Ecologically oriented world systems researchers, ecological unequal exchange researchers, and other macro-structural environmental sociologists also focus little theoretical attention on the role that undemocratic institutions, elite-controlled organizational networks, and capital-state relations play in producing environmental harm;[4] and with some important exceptions (Bunker and Ciccantell 2005; Moore 2000), they also tend to ignore the role that structures of accumulation play in shaping environmental outcomes.[5]

The point I am making is not that macro-structural environmental sociologists fail to discuss these issues, organizations, and networks in their research but rather that as distinct conceptual categories, undemocratic institutions, elite-controlled organizational networks, capital-state relations, and structures of accumulation play little or no role in the theoretical models set forth by *most* macro-structural environmental sociologists.[6] As a result, these issues, organizations, and networks have virtually no *theoretical* significance in macro-structural environmental sociology or environmental sociology more broadly.

This is problematic, I contend, because identifying the kinds of organizational, institutional, and network-based mechanisms that elites use to attain their goals and linking these mechanisms to negative environmental outcomes can illuminate crucial aspects of the structural relationship between human societies and the natural world that are currently missing from much environmental sociology and environmental studies research, as well as from most popular explanations of the environmental crisis. Situating these mechanisms within a theoretical framework is also important because it directs researchers' and environmentalists' attention toward these mechanisms and provides researchers and environmentalists with a theoretical justification for investigating and addressing them. In addition, if I am correct in arguing, as I do later in the chapter, that economic, political, ideological, and military elites must actively create and maintain the undemocratic and network-based mechanisms that they use to achieve their goals, and if they create new mechanisms (or alter old ones) in response to changes in their physical, social, political, and economic environments, then (a) the form that these mechanisms take is likely to vary over time and (b) environmental activists can potentially alter or weaken these mechanisms because the existence of these mechanisms and the form that they take are not set in stone or predetermined by capitalism's functional requirements.

As a result, developing a theoretical model that conceptualizes undemocratic institutions and elite-controlled organizations and networks as *types* of mechanisms that are critical to elite goal attainment, environmental degradation, and human suffering is potentially important for both researchers and environmentalists because it will (a) allow researchers to use a single theoretical model to examine environmental degradation at multiple historical junctures,[7] (b) provide an explanation for why the specific mechanisms responsible for environmental degradation are likely to vary over time and across societies, and (c) place seemingly unrelated mechanisms (such as commodity chains, free trade agreements, military power, and policy planning networks) into a common theoretical framework that holds that these mechanisms play a similar role in harming the environment wherever and whenever they are found, making it easier for environmentalists and researchers to isolate the mechanisms that most critically shape environmental outcomes and identify the points in the social structure that are most in need of reform or abolition.

Thus, in the next section of the chapter, I set forth a new theoretical model, the inequality, democracy, and the environment (IDE) model, that focuses on the role that organizational, institutional, and network-based *inequality* and *undemocratic* and elite-controlled decision making play in producing social and environmental harm (hence the name of the theoretical model). However, before proceeding, I want to stress that I view the IDE model as an important complement to the treadmill of production model, environmental WST, ecological unequal exchange theory, and other macro-structural environmental sociology theories because it focuses on the decision making and enforcement structures that elites use to define and achieve goals and outcomes that are highlighted in many instances by these other theories. In other words, while the IDE model is unique within environmental sociology and while it fills a number of important gaps in the macro-structural environmental sociology literature, I do not view it as challenging other macro-structural environmental sociology theories in any fundamental way.

The one exception to this concerns ecological modernization theory, a putatively macro-structural theory that holds that the dominant organizations, institutions, and networks in wealthy capitalist societies are increasingly incorporating a pro-environmental rationality and logic into their ethical and decision making structures such that (a) these societies are becoming ever more environmentally friendly and (b) further institutional modernization, including increased economic and technological growth and development, represent the best path out of the environmental crisis (Mol 1995, 1997). I spend very little time in this book discussing ecological modernization theory because an extensive body of theory and research effectively refutes its key theoretical claims (for instance, see Foster 2012; Gould et al. 2004, 2008; Jorgenson and Clark 2009; J. Rice 2009; York and Rosa 2003). Nevertheless, the argument and evidence presented in this book demonstrate unequivocally that rather than adopting a pro-environmental rationality and logic, the most powerful organizations, institutions, and networks in the United States and the world work actively to achieve self-interested goals that routinely and severely harm people, societies, and the environment. As a result, and contrary to the predictions of ecological modernization theory, the rationality and logic of greed and power are much more central to the operation of the world's dominant organizations, institutions, and networks than is any pro-environmental rationality or logic. Indeed, no evidence exists to support the claim that the world's dominant institutions have adopted a pro-environmental rationality or logic in any meaningful way at all.[8]

Inequality, Democracy, and the Environment

It is my contention that before environmentalists and environmental social scientists can fully describe the structural relationship that exists between humans and the natural world, they must accomplish three important tasks (I am not arguing that these are the only important tasks they must accomplish). First, they must explain how elites, in attempting to achieve their goals, use and create organizations, institutions, networks, and social structures that simultaneously produce environmental degradation and limit pro-environmental attitudes, values, beliefs, and behavior. Second, they must develop midrange theories and case studies that link the mechanisms elites use to achieve their goals to environmental outcomes such as climate change, increased pesticide use, and decreased soil fertility. Third, they must develop a theoretical link between organizational, institutional, and network-based inequality (OINB inequality) and environmental degradation.

If these assertions are valid, several important questions must be answered, including the following. Who are the elite? What goals do elites want to achieve? What kinds of mechanisms do elites use to achieve their goals? How and why do these mechanisms negatively affect the environment? And how and why are environmental degradation and OINB inequality linked to each other?

Following Michael Mann (1986), I define as elites those individuals who are positioned most advantageously in their society's (or the world's) four most important power networks: economic power networks, political power networks, military power networks, and ideological power networks. Occupying positions of great influence and authority within these networks gives these individuals a greater capacity than others to direct and control these networks, the valued resources that are distributed through these networks, and the people who participate in these networks. As a result, these individuals are better able than non-elites to achieve their goals in the face of resistance from others.[9]

This definition of elites differs from that employed by treadmill theorists, ecologically oriented world system researchers, and ecological unequal exchange researchers, who rely primarily on a Marxist definition of elites as capitalists whose main goal is capital accumulation. Nevertheless, I strongly agree with world systems, treadmill, and ecological unequal exchange researchers that in capitalist societies, capital accumulation is the main goal of economic elites and one of the main goals of political and military elites, both of whom need tax revenues generated from capitalist activity to maintain their power and legitimacy (Schnaiberg and Gould 2000). Political and military elites do, of course, have other important goals. For example, political elites are generally interested in expanding the state's bureaucratic and geopolitical power, and military elites want to increase military budgets and maintain military control over oil, which is vital not only to economic expansion but to military and geopolitical power as well.

However, in this book, I focus most of my attention on capital accumulation, not only because it is critically important to the wealth, power, and legitimacy of economic, political, *and* military elites but also because of the central role it plays in most macro-structural environmental sociology research, because the goals adopted by political elites and governments are often strongly shaped by what capitalist elites want (see Domhoff 1990 as well as chapters 3–6 in this book), and because I focus in this book on events and outcomes in the *capitalist* world system over the past 70 years.

Capitalists and core nation governments have several options available to them in their attempts to foster capital accumulation. Among other things, they can lower trade barriers between nations, expand markets, alter property rights laws, reduce government regulations, decrease labor costs, improve their terms of trade, and minimize the costs of extracting and transporting vital raw materials. However, in order to accomplish any of these tasks, capitalists and core nation governments must be able to achieve their goals in the face of resistance from others, which implies the existence of inequality, or the unequal distribution of economic, political, military, geopolitical, and ideological power. The ability of economic and noneconomic elites to achieve *other* important goals is likewise predicated on unequal access to economic, political, military, geopolitical, and ideological power.

As a result, if elites, in pursuing their goals, regularly and seriously degrade the environment, then any structural explanation of environmental degradation must develop a theoretical link between inequality, elite power, and the natural world (this is true even if environmental degradation is an *unintended* consequence of elite activity, which it likely is in many cases).[10] James Boyce (2002), an economist at the University of Massachusetts, begins to develop such a link in his book *The Political Economy of the Environment.*

Boyce argues that environmental degradation typically involves winners and losers. Winners are able to shift environmental costs onto losers because the losers have not yet been born, are not aware of the costs imposed on them due to lack of information, or are aware of the costs imposed on them but lack the power to prevent the winners from achieving their goals. Boyce's argument has two important implications. First, it suggests that the ability to engage in environmentally destructive activity is predicated in large part on the ability to shift environmental costs onto others[11] and that a critical factor allowing this to occur is power disparities, or inequality, between groups. Second, it suggests that inequality and environmental degradation are inextricably linked to each other. Thus, overcoming environmental degradation necessarily involves overcoming inequality.[12]

Boyce provides a solid foundation on which to develop a structural theory of OINB inequality, elite power, and environmental degradation. However, the ability of elites to shift environmental costs onto others is not the only way in which elite power and OINB inequality lead to environmental degradation. Moreover, Boyce does not discuss the organizational, institutional, and network-based mechanisms that elites use to accomplish their goals and shift costs onto others.

Expanding on Boyce's argument, the inequality, democracy, and the environment model holds that OINB inequality (in the economic, political, military, geopolitical, and ideological spheres) is a fundamental source of environmental degradation because it (a) allows a small number of people and organizations to monopolize decision making power; (b) allows more powerful individuals, groups, and organizations to shift environmental costs onto less powerful individuals, groups, and organizations; (c) inhibits the development and/or dissemination of environmental knowledge, attitudes, values, and beliefs, at least when this knowledge and these attitudes, values, and beliefs conflict with elites' perceived interests; (d) restricts the ability of non-elites to behave in environmentally sustainable ways by limiting the choices and shaping the incentives available to them; (e) allows elites to successfully frame what is and is not considered to be pro-environmental behavior, pro-environmental policy, and pro-environmental development (McCright and Dunlap 2003; Goldman 2005); and (f) allows elites to divert the public's attention away from what they are doing so that their actions will not be scrutinized, questioned, or challenged (Freudenburg and Alario 2007).

OINB inequality does not give rise to these outcomes automatically, however. Instead, as Domhoff (1990, 2002) has demonstrated, elites have to work actively to achieve their goals, which means that any structural theory of environmental degradation must explain how OINB inequality is maintained, how elites exert power over non-elites, and how elites accomplish the tasks listed in the preceding paragraph.[13] In order to develop such an explanation, I focus initially on Michael Mann's (1986) research on the organizational and network bases of social power. Mann argues that there are four types of social power: ideological power, economic power, political power, and military power. None of these sources of power is ultimately determinative, none of them can be reduced to any of the others, and although they tend to converge as they become more institutionalized, they never become fully institutionalized and they never represent a unitary source of social power.

What most distinguishes Mann's theory from other theories of power and stratification (inequality) is his argument that these four sources of social power are both "overlapping and intersecting sociospatial networks" and "organizations, institutional means of attaining human goals" (1986, 1–2). In emphasizing the organizational and network bases of social power, Mann directs readers' attention away from abstract structural causes to the actual organizational and network-based mechanisms that make institutionalized stratification and elite goal attainment possible.

Mann's theory thus suggests that in order to link elites to environmental degradation, environmental social scientists must identify the organizations and networks that provide the institutional basis for elite power and trace the "flow" of elites and power through these organizations and networks. In other words, environmental social scientists must study the specific organizational, institutional, and network-based mechanisms that elites use to achieve their goals because elites could not achieve their goals without these mechanisms and because these mechanisms form the basis of and critically shape the large-scale social structures that harm the environment, make capital accumulation possible, and limit the behavioral choices available to non-elites.

For the purposes of this book, I define elite-controlled mechanisms as organizational, institutional, and network-based tools that provide elites with the means to *define their interests*, *make decisions*, and *achieve their goals* in the face of resistance from others. I further argue that these mechanisms tend to be housed in *undemocratic* and *elite-controlled* organizations, institutions, and networks because elites are greatly outnumbered by non-elites and thus cannot afford to provide non-elites with democratic access to organizational, institutional, and network-based mechanisms that would allow them to fundamentally shape economic, political, geopolitical, military, and environmental policy.

The IDE model thus holds that environmental degradation is to a significant degree the product of OINB inequality, which provides economic, political, military, geopolitical, and ideological elites with the means to create and control undemocratic organizational, institutional, and network-based mechanisms through which they are able to monopolize decision making power, shift environmental and non-environmental costs onto others, frame what is and is not considered to be good for the environment, divert the public's attention away from what they are doing, and shape individuals' knowledge, attitudes, values, beliefs, and behavior. These undemocratic mechanisms produce severe human, social, and environmental harm, I argue, because they provide elites with the means to achieve goals that are often environmentally, socially, and humanly destructive and because these mechanisms are sometimes environmentally, socially, and humanly destructive in and of themselves, as is the case with military power and state-sponsored armed violence.

An important implication of this that I have not yet discussed but that I highlight in chapter 5 is that the organizations, institutions, and networks within which elite-controlled mechanisms are housed tend to be violent in at least one of three ways. They either (a) use or threaten to use armed violence to achieve their goals, (b) achieve their goals by relying at least in part on armed violence carried out by other organizations, institutions, or networks, or (c) severely harm, either physically, emotionally, or psychologically, individuals, communities, societies, and/or the natural world.[14]

The undemocratic and elite-controlled mechanisms highlighted by my theoretical model also form the basis of and critically shape the large-scale social structures and structures of accumulation that I previously argued negatively affect the environment and (in capitalist societies) make capital accumulation possible. These mechanisms, and the decisions that elites make in the organizations, institutions, and networks they control, also create a body of accepted knowledge and a set of social structures, cultural values, ideological justifications, and behavioral constraints that drastically limit the agency of non-elites. As a result, non-elites are able to take very few actions and make very few decisions that do not harm themselves, the environment, and/or others, and are thus forced to engage in behavior that is often humanly, socially, and environmentally destructive.

Because the cultural, ideological, and knowledge-based constraints that elite-controlled organizations, institutions, and networks impose on non-elites are so pervasive, it is also difficult for non-elites to fully understand how their social system works or to imagine alternative structural arrangements that would promote a better, more humane and sustainable world; and because elites dominate and control the world's most powerful organizations, institutions, and networks, including the military and police, it is also very difficult (though not impossible) for non-elites to effectively struggle for positive social and environmental change.

The IDE model thus holds that non-elites usually have little choice but to participate in local, national, and global social systems that harm both them and the environment, not because they are stupid or lazy or because they would not benefit from radical social change but, to paraphrase Mann (1986, 7), because they are *organizationally outflanked* by elites and thus have little power to shape the world in which they live. Finally, because political, economic, ideological, and military power is held by elites and exercised through organizations, institutions, and networks that they control, the IDE model also places overall responsibility for the world's social and environmental crises on elites (rather than on consumers or average people), and it attributes this responsibility to elites regardless of whether elites actually intend to harm the people, societies, and environments that they harm.[15]

Evaluating the Theoretical Model

In the chapters that follow, I present a series of case studies that (a) illustrate the validity of the IDE model across a variety of empirical settings and (b) demonstrate that social scientists and environmentalists cannot adequately explain the causes of environmental degradation without examining either the link between OINB inequality and environmental degradation or the institutional, organizational, and network-based mechanisms through which elites exert power. These case studies also test my argument that undemocratic organizations, institutions, and networks play a key role in harming non-elites and the environment because they provide elites with the means to monopolize decision making power, shift environmental and non-environmental costs onto others, divert the public's attention away from what they are doing, frame what is and is not considered to be good for the environment, and shape individuals' knowledge, attitudes, values, beliefs, and behavior.

In doing these things, the case studies also demonstrate that structural factors ignored or downplayed by the treadmill model, environmental WST, and ecological unequal exchange theory but highlighted by the IDE model play an important role in shaping environmental and social outcomes, thereby helping to clarify how the IDE model both differs from and contributes to these other theories.

Finally, the case studies test my claim that a lack of democracy in elite-controlled organizations, institutions, and networks plays a critical role in producing social and environmental harm. I define democracy in both narrow and broad terms. On the one hand, I define democracy fairly narrowly as a political system in which every individual, group, and organization has a relatively equal opportunity to influence political decisions and public policy and no individual, group, or organization has a highly disproportionate opportunity to do so. In this narrow sense, democracy is about political decision making processes and the organizations, institutions, and networks that provide individuals, groups, and other nonstate actors with equal or unequal opportunities to influence the government.

On the other hand, democratic and undemocratic decision making can also occur in socially, politically, and environmentally consequential nonstate organizations, institutions, and networks such as international trade and finance institutions (e.g., the World Bank, IMF, WTO, and NAFTA), commodity chains, and policy planning networks. As I demonstrate in the next several chapters, international trade and finance institutions, commodity chains, and policy planning networks are all organizations, institutions, or networks used by elites to define their interests, make and enforce decisions, and achieve their goals in the face of resistance from others. We can thus ask of these and other similar organizations, institutions, and networks whether they provide elites with disproportionate decision making authority that allows elites to achieve goals that severely harm individuals, groups, and societies that have little or no opportunity to influence the decisions made within these organizations, institutions, and networks.

In order to test my theoretical claims, demonstrate the validity and widespread applicability of the IDE model, and more clearly differentiate the IDE model from other macro-structural environmental theories, the case studies I present in the following chapters examine a wide range of environmental outcomes and undemocratic, elite-controlled mechanisms. For example, the case studies in chapter 3 highlight a variety of mechanisms employed by the World Bank and IMF that simultaneously benefit elites and harm the environment; the case studies in chapter 4 demonstrate how economic and political elites use agricultural commodity chains, liberalized agricultural trade, the WTO's Agreement on Agriculture, and U.S.-based policy planning networks to achieve U.S. and agriculture industry goals; chapter 5 demonstrates that armed violence, developing nation debt, and World Bank lending policy are mutually reinforcing elite-controlled mechanisms that provide corporations with the means to exploit developing nations' natural resource wealth; and chapter 6 examines the role that U.S. policy planning networks, White House links to specific organizations within these policy planning networks, and restricted decision making played in the formation of the George W. Bush administration's energy and military policies.

I selected these case studies for two important reasons. First, they highlight a set of very powerful elite-controlled mechanisms that have had particularly severe human, social, *and* environmental consequences. Second, they each highlight different, though sometimes overlapping, sets of mechanisms that elites use to pursue their goals. This is important because examining different elite-controlled mechanisms in different case studies allows me to show how seemingly unrelated mechanisms contribute to elite goal attainment, environmental degradation, and social harm in theoretically similar ways. Conversely, examining the same type of mechanism (e.g., policy planning networks) in different case studies allows me to demonstrate how the same type of mechanism can operate both similarly and differently in different contexts. Finally, presenting multiple case studies in each chapter allows me to test multiple theoretical claims both within and across chapters and environmental issue areas. For instance, the case studies presented in chapter 3 test all six of the theoretical claims I make about why elite-controlled organizations, institutions, and networks harm individuals, societies, and the environment, while chapters 4–6 each test between three and five of these claims (see table 2.1). As a result, each of these claims is evaluated in multiple empirical settings, thereby providing strong support for the IDE model and demonstrating that the social structural relationships emphasized by the model are not unique to a single empirical case.[16]

TABLE 2.1. Theoretical Predictions Tested in Each Chapter

	Predictions: Control of undemocratic organizations, institutions, and networks provides elites with the means to:					
Chapter	(a) Monopolize decision making power	(b) Shift environmental and non-environmental costs onto others	(c) Shape environmental knowledge, attitudes, values, and beliefs	(d) Frame what is and is not considered to be pro-environmental behavior, policy, and development	(e) Shape environmental behavior by restricting choices and shaping incentives	(f) Divert public attention away from what elites are doing
3	X	X	X	X	X	X
4	X	X			X	
5	X	X	X	X		X
6	X	X			X	X

Before presenting the case studies, there are two issues I need to address. First, because the case studies cover so much empirical ground, I do at times make arguments that are broadly similar to those made by other researchers. What differentiates my arguments and case studies from those of other scholars is that regardless of the specific arguments I make in any single case study or chapter, the goal of every case study and chapter is to test my theoretical model and demonstrate that regardless of their form or function elite-controlled organizations, institutions, and networks play a key role in harming people, communities, societies, and the environment. As a result, even when my specific arguments are similar to those made by other researchers, my goal in presenting these arguments is different, as are the theoretical conclusions I draw from both these arguments and the evidence I present. I am also more likely than other scholars whose arguments overlap mine to emphasize the organizational, institutional, and network bases of elite power.

Second, earlier in the chapter I noted that properly theorizing capital-state relations and the policy making process is of critical importance if we are to fully understand the macro-structural roots of the current environmental crisis, the role that economic and political elites play in harming the environment, and the mechanisms that elites use to achieve their goals. In the next section of the chapter, I therefore briefly describe power structure theory, which holds that in the United States capitalists use a very specific set of organizational networks to dominate the policy making process (Burris 2005; Domhoff 1990, 2002). I highlight power structure theory rather than some other theory of capital-state relations because several of my case studies focus on policy making in the United States and because power structure theory is more consistent with the basic tenets of the IDE model than are any of the other theories available to me.

It is important to note, however, that the organizational networks highlighted by power structure researchers are just one type of elite-controlled mechanism that can potentially harm people, communities, and the environment. Moreover, these networks do not operate in all societies or historical time periods, and they are only one of several elite-controlled mechanisms that I highlight in the book. In other words, although they currently play a very important role in producing global environmental degradation, they are not the only important mechanism elites use (or have used) to achieve their goals, and in highlighting them, I am not narrowing the focus of the IDE model to this one mechanism. Finally, I recognize that it is a bit awkward to end the chapter with a discussion of power structure theory. However, it is critically important that readers understand the basic outlines of this theory, and it makes more sense to describe the theory here than earlier or later in the book.

Power Structure Theory

According to power structure theory, U.S. policy is influenced disproportionately by a group called the power elite (Burris 2005; Dreiling 2001; Mizruchi 1992, 1996), which G. William Domhoff (1990, 2002), one of the subfield's central researchers, defines as the economic and political leadership of the upper class and the most prominent members of the corporate community. Domhoff (2002) identifies as leaders of the upper class and corporate community individuals who serve as directors on multiple corporate boards of this nation's most powerful corporations, banks, and law firms. Sitting on multiple corporate boards gives these individuals a class-wide perspective on public issues (Domhoff 2002) and helps tie major corporations into a common "corporate board interlock" network that can be mobilized for political action (Dreiling 2001; Mizruchi 1992, 1996).

The power elite exert influence over public policy, power structure researchers argue, by lobbying political officials and influencing the selection of political candidates. However, their most important source of influence over public policy is a set of linked organizations that power structure researchers call the policy planning network. This network consists of policy discussion groups such as the Council on Foreign Relations, the Committee for Economic Development, and the Trilateral Commission; foundations such as the Rockefeller and Ford Foundations; and think tanks such as the Brookings Institution and the American Enterprise Institute. The power elite fund and serve as directors and trustees of these organizations, power structure researchers hold, thereby tying the different organizations within the network to each other and to the power elite and corporate community (Domhoff 2002).

According to Domhoff (2002), social, economic, and political problems that are initially discussed in corporate boardrooms are often sent to think tanks, where policy experts employed by the think tanks develop possible policy solutions for these problems as well as for problems that the think tanks believe should be of concern to the upper class and corporate community. These solutions are then sent to policy discussion groups, where their relative merits are discussed and debated by members of the power elite and policy planning network. The policy planning network thus provides members of the power elite with the expertise they need to identify and achieve their policy goals, with this expertise essentially being purchased by the power elite, whose money is channeled to think tanks and policy discussion groups by the foundations that also make up the network.

Power structure researchers further argue that the policy planning network and power elite are linked to the federal government in various ways. Network policy reports are read by government officials and their staffs, and members of the network and the power elite testify before Congress, are regular members of unpaid executive branch advisory committees, are prominent on presidential commissions, are appointed by the president to Cabinet posts and other important executive branch positions, and serve as informal advisers to the president during foreign policy crises. Members of the power elite also belong to organizations such as the Business Council and the Business Roundtable that are linked to the policy planning network and have close formal and informal ties to the executive and legislative branches of the federal government. These network connections between the power elite, policy planning network, and the state not only provide the power elite and policy planning network with disproportionate access to the White House and Congress, they also provide the power elite and policy planning network with disproportionate influence over U.S. foreign and domestic policy (Burris 1992, 2005; Domhoff 1990, 2002; Dreiling 2001; Mizruchi 1992, 1996), thereby undermining democracy in the United States.

Moreover, while the people who work in the policy planning network are not the puppets of the power elite, they are in an indirect fashion their employees, and when they interact with the government in this capacity or enter the government as former network employees, they are acting either as representatives of the power elite or as people whose ideas the power elite funded and therefore helped create. As a result, when members or former members of the policy planning network interact with U.S. officials or enter the U.S. government, they provide the power elite with an indirect connection to, and potential influence over, U.S. officials, particularly when these former network employees become high-level decision makers within the White House and executive branch.

Power structure theorists recognize, of course, that the power elite do not win every public policy debate, that the solutions they and the policy planning network devise are not always well designed or successfully implemented, and that the power elite and capitalist class are internally divided in several important ways. Nevertheless, power structure theory provides researchers with an organization-, network-, and mechanism-based account of policy making in the United States that can be used to better understand the relationship between OINB inequality, democracy, and environmental degradation.

I turn now to an examination of the social and environmental consequences of globalization, in which I highlight a set of elite-controlled mechanisms that have played a key role not only in promoting globalization but also in harming people, societies, and the environment.

3

The World Bank, the International Monetary Fund, and the Environment

Over the past several decades, there has been a strong push by the world's wealthiest nations to more closely integrate developing nations into the global economy by privatizing their state run businesses, increasing their exports, weakening and eliminating their barriers to international trade and investment (trade and investment liberalization), and reducing the role their governments play in their economies. This movement toward greater global economic integration, generally referred to as globalization,[1] is described by many pundits, politicians, and members of the media as being an inevitable and inexorable force that represents the only viable path for overcoming world poverty and ensuring national and global economic growth (Steger 2009; Stiglitz 2002). Others argue, however, that global economic integration, at least in its current form, retards economic growth in developing nations while increasing inequality and poverty around the world. From this perspective, global economic integration represents one of the ongoing political triumphs of neoliberalism, which can be defined as (a) an economic and political philosophy that holds that economic and social policy are best determined in the marketplace, (b) an ideology that is used to justify a set of free market policies that benefit specific groups of elites to the detriment of developing nations and the world's poor, working, and middle classes, (c) a political movement that has successfully struggled within the U.S. and globally to achieve a specific set of free market policy goals, and (d) a set of organizations and institutions through which free market policy goals such as trade liberalization and reducing government involvement in the economy are achieved.[2]

Of course, global economic integration cannot be both inevitable and inexorable and the product of a political movement that has had to struggle to achieve its goals; nor can it both alleviate and reinforce world poverty. Thus, one of the main goals of this chapter is to deter-

mine whether global economic integration is inevitable and inexorable
and whether it exacerbates or alleviates world poverty. The second, but
more important, goal of the chapter is to determine whether the World
Bank and the International Monetary Fund (IMF), two powerful neoliberal institutions that have played key roles in promoting global economic integration, are elite-controlled and undemocratic, whether they
are dominated by actors operating in or through other elite-controlled
organizations, institutions, and networks, and whether the policies they
promote simultaneously achieve elite goals and produce social and environmental harm in the developing nations where much of their work
is directed. Because I end up concluding that each of these assertions is
true, and because the evidence presented in this chapter demonstrates
that the World Bank and IMF both work hard to achieve specific neoliberal goals, I am able to demonstrate two important points: first, that
these institutions play a theoretically similar role in producing social
and environmental harm as do those elite-controlled organizations, institutions, and networks that I examine later in the book and, second,
that global economic integration is not the product of inevitable and
inexorable economic forces but of specific decisions made by powerful
actors in a set of undemocratic institutions that have a very specific vision of what globalization should look like.

In addition to demonstrating these points, I use the evidence presented in this chapter to test the six hypotheses I set forth in chapter 2.
These hypotheses hold that control of undemocratic organizations, institutions, and networks provides elites with the means to (a) monopolize
decision making power, (b) shift environmental and non-environmental
costs onto others, (c) inhibit the development and/or dissemination of
environmental knowledge, attitudes, values, and beliefs, (d) frame what
is and is not considered to be pro-environmental behavior, policy, and
development, (e) divert public attention away from what elites are doing
so that their actions will not be scrutinized, questioned, or challenged,
and (f) shape environmental behavior by restricting the choices and
shaping the incentives available to non-elites. As we shall see, all six hypotheses are supported by the evidence presented in this chapter.

The chapter is organized as follows. I begin by briefly describing the
World Bank and IMF and then discuss the severe social, political, economic, and environmental consequences of their structural adjustment

policies and of the World Bank's project-based lending and knowledge production policies. I then present evidence that World Bank and IMF policy are shaped primarily by the world's most economically powerful nations, in particular, the U.S. I demonstrate that these nations influence the World Bank and IMF through a set of organizational mechanisms that give particular government officials within these nations disproportionate influence over these two institutions, and I show that in the 1980s the neoliberal structural adjustment policies that the U.S. forced the World Bank and IMF to impose on developing nations were shaped in key ways by powerful economic actors within the U.S. I further demonstrate that these powerful American actors used elite-controlled organizations and networks within the U.S. to strongly influence the policy positions the U.S. took at the World Bank and IMF. I thus conclude that in using these undemocratic organizational mechanisms to achieve their goals, a very small group of economic and political elites in the world's wealthiest nations have strongly and negatively influenced social, economic, political, and environmental conditions in a large number of developing nations. This finding is, of course, highly consistent with my theoretical argument.[3]

The World Bank and the International Monetary Fund

The World Bank and IMF were both created near the end of World War II at the Bretton Woods conference in Bretton Woods, New Hampshire.[4] This conference brought together 730 delegates from the U.S. and the 43 nations allied with the U.S. during the war to create a postwar monetary system and system of economic governance that was designed both to ensure U.S. economic and geopolitical dominance in the postwar world (Domhoff 1990) and to prevent the global economy from falling into another economic depression (Peet 2003).[5]

As originally designed, the work of the World Bank and IMF, albeit critically important, was fairly circumscribed. The chief aim of the World Bank was to provide loans to European nations for postwar reconstruction efforts, while the IMF was to regulate member nations' currency exchange rates and lend short-term funds to nations undergoing balance of payment crises (Peet 2003). However, by the end of the 1950s, the World Bank had shifted its focus from European reconstruction to providing developing nations with loans for large-scale economic development projects, primarily in the areas of agriculture, natural resource extraction, infrastructure, and energy generation and distribution; and by the mid-1980s, the IMF and World Bank were both making loans to low- and middle-income countries that required these countries to undertake deep and wide-ranging policy reforms across a broad range of policy areas not considered by the framers of the Bretton Woods agreement, including by the first decade of the 21st century, reforms to these nations' legal codes, financial accounting practices, natural resource policies, and methods of providing services such as clean drinking water (Goldman 2005; G. Harrison 2004; Woods 2000).

Indeed, World Bank and IMF lending practices in the developing world have become so pervasive, and the scope of conditions that developing nations must meet to receive World Bank and IMF loans has become so extensive, that the World Bank and IMF have, for the past 30 years, played a key role in shaping a broad array of policies and outcomes in low- and middle-income countries around the world. Thus, in the following sections of the chapter, I highlight some of the most important ways in which World Bank and IMF lending practices have shaped economic, social, political, and environmental outcomes in developing nations, with a focus on IMF and World Bank structural adjustment lending, World Bank project-based loans, and the knowledge generating structures that the World Bank uses, at least in part, to justify and promote its lending policies.

Structural Adjustment Loans (Policy-Based Lending)

Structural adjustment loans are policy-based loans that the IMF and World Bank make to low- and middle-income countries that are experiencing severe balance of payment problems[6] due to factors such as "large and persistent fiscal deficits; high levels of external and/or public debt; [currency] exchange rates [that the World Bank and IMF believe are] inappropriate[;] . . . [and] sudden and strong increase[s] in the price of key commodities such as food and fuel" (IMF 2008, 1). Structural adjustment loans come with conditions that require debtor nations to adopt specific fiscal, monetary, exchange rate, and structural policies[7] designed to reduce the role of the government in the economy, increase the role of markets and the profit incentive in determining economic outcomes, strengthen private property rights, alter debtor nations' macroeconomic environments, and shift the focus of economic activity away from production for domestic consumption toward production of goods and natural resources for export. These adjustment policies, the IMF and World Bank argue, provide borrower countries with the means to address their balance of payment problems, service their debts, reduce inflation and poverty, and promote economic growth (Peet 2003).

Specific loan conditions vary from one structural adjustment loan to another but generally include policies such as radically reducing government spending, increasing interest rates, liberalizing trade (reducing and removing tariffs and other import and export restrictions), devaluing the national currency, removing restrictions on foreign investment, drastically reducing government regulation of the market, privatizing state enterprises, removing price controls, reducing or abolishing specific government subsidies, improving public sector and corporate governance, reinforcing the rule of law, and revising government regulations that inhibit the extraction of natural resources by foreign-owned companies (Bello et al. 1999; Rieffel 2003; UNCTAD 2005). To ensure compliance with loan conditions, structural adjustment loans are disbursed in installments, or tranches, which allows the World Bank and IMF to punish nations that fail to comply with their loan conditions by withholding future loan disbursements from them.

Although the first structural adjustment loans were granted by the World Bank in 1980 (to Turkey and Kenya), their use did not become widespread until later in the decade when the U.S., the World Bank, and the IMF made them the centerpiece of their strategy to address a severe debt crisis sweeping the developing world at that time. Not surprisingly, many developing nations were reluctant to accept loans that so greatly intruded on their sovereign right to regulate their markets and economies, but given the scale of the crisis in which these nations found themselves, they generally had no choice but to accept structural adjustment loans on the terms set forth by the World Bank and IMF, with 89 countries receiving such loans by 1991 (Bello et al. 1999).

Structural adjustment loans have gone by various names over the years, and their policy focus has changed in several important respects since 1980. Nevertheless, they have been used extensively since the early 1980s to impose neoliberal policies on developing countries experiencing severe debt crises.[8] Moreover, the number of adjustment conditions attached to World Bank and IMF loans has increased, at times quite dramatically, since they were first introduced. For instance, the number of conditions attached to the average IMF structural adjustment loan rose from seven in the early 1980s to 16 in 1997 before dropping to 8.8 in 1998 (Bird 2001), most likely as part of a temporary response to intense criticism of IMF loan conditionality following the 1997 East Asian financial crisis (Vreeland 2007). Consistent with this argument, Eurodad (2006) found that between 2002 and 2006 the average number of structural conditions associated with IMF Poverty Reduction and Growth Facility loans (a type of structural adjustment loan) had risen again to 11; and because Eurodad did not include macroeconomic conditions, such as quantitative targets on fiscal deficits and domestic credit creation, in its study, its findings actually underrepresent IMF loan conditionality in this period.

Eurodad (2006) also examined structural adjustment conditionality in 20 developing nations that received World Bank loans between 2003 and 2005. Three of the World Bank loans included in Eurodad's study had more than 100 conditions and 14 had more than 50 conditions, with an average of 15 binding and 52 nonbinding conditions per World Bank loan (at the time, 77% of developing nation loan recipients thought they had to adopt nonbinding loan conditions).

Eurodad (2006) also found that although IMF and World Bank structural adjustment loans still required borrower countries to enact "traditional" adjustment policies such as trade liberalization and privatizing government run businesses, there was at the time of the study a new emphasis on privatizing the telecommunications, energy, and banking sectors and on enacting anticorruption, civil service, legal, judicial, tax, and public finance management reforms. And while the IMF and World Bank argue that they now provide debtor nations with more influence, or "ownership," over what is included in structural adjustment loan agreements, a 2005 World Bank survey found that "50% of recipient countries felt that the World Bank introduced elements that were not part of the country's [original] program into their loan conditions" (quoted in Eurodad 2006, 14). Moreover, in a study of structural adjustment in Africa, Graham Harrison (2004, 88) demonstrates that although Uganda, Tanzania, and Mozambique each drafted plans for reducing poverty and promoting economic growth that were supposed to form the basis for subsequent IMF and World Bank structural adjustment loans, these plans still had to be approved by the IMF and World Bank. This, of course, gave the IMF and World Bank final say over what these highly indebted countries were able to propose and calls into question the idea that these and other borrower nations, all of which are highly dependent on World Bank and IMF financing (see the following section), "own" their economic adjustment plans. Indeed, the fact that debtor nations are highly dependent on the World Bank and IMF strongly suggests that these nations are careful to include in their plans exactly what they believe these global institutions want them to include.

Why Do Developing Nations Accept Structural Adjustment?

Given the intrusive nature of World Bank and IMF structural adjustment loans and the fact that many nations do not feel ownership over them, why is it that so many developing nations accept these loans? The most likely reason is that they have little choice. Nations that seek out structural adjustment loans generally do so because they are experiencing a severe economic crisis that requires outside funding that they cannot obtain without IMF and World Bank support. This support comes not only in the form of World Bank and IMF loans but also from the seal of approval that debtor nations receive from the IMF and World Bank, which provides them with access to other sources of funding that are otherwise unavailable to them (UNCTAD 2000). As a result, accepting structural adjustment loan conditions is not a decision nations freely make but rather one that is forced on them by the World Bank, IMF, and circumstance, in a situation where the IMF and World Bank could likely offer more attractive and economically effective alternatives (see the following sections) that would impinge less on debtor nations' sovereign rights.

Another possible reason that countries accept structural adjustment loans is that they may not have to meet the conditions laid out in the loans or do not believe that they have to meet them. There is some support for this argument. Research shows, for example, that the degree to which different nations comply with loan conditions can vary widely and that developing nations that are aligned with the U.S. or other wealthy nations tend to be punished less severely for failing to meet loan conditions than are other developing nations, with punishment generally defined as the withholding of previously promised World Bank and IMF funds (Kilby 2009; Vreeland 2007).

However, many developing nations *are* punished for failing to comply with loan conditions, and this is particularly true of those that are not aligned with a powerful patron such as the U.S. Moreover, the United Nations Conference on Trade and Development (UNCTAD 2000) found that in a sample of 17 least developed countries that received IMF structural adjustment loans between 1986 and 1994, only 20 of the 34 interruptions to IMF funding that these nations experienced were due to a failure to meet the requirements of the loan agreement. And of these 20 failures, 15 resulted from external shocks to the economy, natural disasters, or social unrest (UNCTAD 2000). Some of the 20 failures were also due, at least in part, to poor economic forecasting on the part of the IMF or to a lack of sufficient funding for the required policy reforms; and as the authors of the UNCTAD report note, it is difficult to blame debtor nations for any of these problems. In most of these 34 cases of loan interruption, then, there is no evidence that debtor nations tried to avoid meeting their loan obligations.

Further contradicting the argument that debtor nations do not meet or do not believe they have to meet their structural adjustment loan obligations, UNCTAD (2000, 103–106) also found that by the mid-1990s developing nations that were engaged in IMF structural adjustment programs had achieved, on average, high or near-high scores (relative to "a specified notion of best practices") on trade liberalization reform, exchange rate reform, and price and marketing reform, and moderate scores on financial sector reform and the reform of the public enterprise sector. The evidence thus suggests that even though structural adjustment loan conditions are not always met, developing countries have made large strides toward adopting the kinds of neoliberal policies promoted by the IMF and World Bank. The evidence further suggests that regardless of whether developing nations want to "adjust," they take the conditions laid out in structural adjustment loans seriously.

*The Economic, Social, and Environmental Consequences of
Structural Adjustment*

Another possible explanation for why so many developing nations have
entered into structural adjustment loan agreements is that structural
adjustment may, in fact, promote economic growth and reduce poverty,
inflation, national debt, and balance of payment deficits. This, unfortunately, is not the case, suggesting yet again that developing nations sign
structural adjustment loan agreements not because they want to do so
but because they are forced to do so.

Indeed, a broad body of quantitative research that examines the
economic effects of IMF structural adjustment loans across multiple
countries shows that these loans have no effect on inflation, increase
income inequality, *reduce* economic growth, particularly among nations
that have completed their IMF programs, and in democracies, decrease
government spending on health and education (these studies are summarized in Vreeland 2007; also see UNCTAD 2000).[9]

Countries that have undergone structural adjustment have also seen
their debt levels increase substantially over time, making them more,
rather than less, dependent on the IMF and World Bank (Toussaint
2005; UNCTAD 2000). For instance, between 1980 and 2002, Latin
American and Caribbean nations' debt increased from $257.4 billion to
$789.4 billion, and by 2002 African nations had repaid four dollars for
every dollar of debt they owed in 1980, while still owing four dollars for
each dollar of their original debt (Toussaint 2005, 277). Moreover, between 1998 and 2002, "the governments of sub-Saharan Africa received
$34.83 billion in fresh loans . . . [but] repaid $49.27 billion on previous
loans, . . . [thus] transferr[ing] [nearly] $15 billion . . . to creditors in the
North. Each year, sub-Saharan Africa [also] pays more in debt servicing
than the total of all health and education budgets for the entire region"
(Toussaint 2005, 277–278).

As if all this were not bad enough, a 2004 study conducted by SAP-
RIN[10] (2004) that qualitatively examined the effects of World Bank
structural adjustment lending on social, political, economic, and environmental outcomes in nine nations (Bangladesh, Ecuador, Ghana,
Hungary, Mexico, the Philippines, Zimbabwe, El Salvador, and Uganda)
found that World Bank structural adjustment programs have a wide

range of profoundly negative consequences for adjusted nations. Summarizing some of this study's most important findings, Peet notes that in the nine nations included in the study, structural adjustment "devastated local industries, especially small and medium-sized enterprises providing most national employment, . . . undermined the viability of small farms, weakened food security, . . . damaged the natural environment, . . . undermined the position of workers, causing employment to drop, real wages to fall and workers' rights to weaken, . . . disproportionately reduced the poor's access to affordable services [including utility services, schools, and health care], . . . [and disproportionately harmed] women" (2003, 143).[11] Moreover, these outcomes have been associated with structural adjustment in other nations too (Bello et al. 1999; Naiman and Watkins 1999; Toussaint 2005).

The SAPRIN study also highlights several critical environmental problems associated with structural adjustment. For instance, in the agriculture sector, structural adjustment programs generally favor large-scale export production over smallholder farming, which in the nations included in the SAPRIN study resulted in a decrease in the incomes and viability of smallholder farms. This, in turn, forced smallholder farmers to intensify agricultural production and expand their farming activities to marginal lands, activities that have been associated in other research with increased soil erosion, reduced soil quality, and the degradation of local waterways. At the same time, to meet the growing demand for agricultural exports, wealthier farmers turned increasingly to monoculture farming (single-crop farming), which reduced soil fertility, drew down water resources, eroded the soil, relied on heavy fertilizer and pesticide applications, and polluted farmland and waterways. In addition, as agricultural export production increased, production for local consumption decreased, thereby increasing local food prices and hunger (SAPRIN 2004).

SAPRIN (2004) also documented the economic, social, and environmental consequences of mining sector structural adjustment reform in two regions of the Philippines (Didipio and Manicani Island) and the Tarkwa region of Ghana. (Because World Bank structural adjustment loans forced many developing nations to implement mining sector reforms in the 1980s and 1990s [see chapter 5], it is likely that SAPRIN's findings for Ghana and the Philippines hold for many other developing nations too.)

In Ghana, for example, mining sector reform failed to produce economic growth, despite helping to drastically increase mining sector exports, because the reform policies promoted by the World Bank severely limited the royalties and taxes that mining companies had to pay the Ghanaian government, thereby allowing these companies to take most of their profits out of the country.[12] Mining activities associated with mining sector reform also contributed to the economic and residential displacement of tens of thousands of people, due both to forced resettlement and the removal of large tracts of land from the agricultural sector. Because the number of displaced farmers exceeded the number of new jobs created in the mining sector, unemployment in the region also rose at the same time as the cost of living near the mines increased (SAPRIN 2004). Mining sector reforms in Ghana and the Philippines also

> allowed large-scale mining to expand without effective environmental controls, thereby polluting local and regional environments and degrading sensitive, biologically rich zones. Mechanisms to conduct environmental impact assessments exist[ed] in both Ghana and the Philippines, but adjustment measures [such as reduced government spending] . . . left the governments of those countries with little capacity to enforce this requirement effectively or to ensure compliance with environmental quality standards. As a result, mines . . . often lowered water tables, diverted watercourses, and caused water pollution through the use of chemicals and the unleashing of heavy metals. The widespread removal of trees and vegetation . . . also resulted in soil erosion and decreased soil fertility, which . . . made land unsuitable for agricultural purposes. In addition, mining operations . . . destroyed traditional, ecologically sound systems of shifting cultivation by reducing fallow periods as a result of the reduction in land available for cultivation. Furthermore, mining and related activities . . . contributed to air pollution through the release of particulate matter and emissions of black smoke. (SAPRIN 2004, 171)

Like the SAPRIN report, other studies have linked a variety of environmental problems to structural adjustment. For instance, a 1996 World Wildlife Fund study (Reed 1996) that examined the social, economic, and environmental effects of structural adjustment in Cameroon, Mali, Tanzania, Zambia, El Salvador, Jamaica, Venezuela, Pakistan, and Vietnam found that export growth promotion, one of the key elements of most structural adjustment programs, generally resulted in the increased extraction of natural resources (minerals, timber, precious metals, oil, etc.) from these nations. This not only helped to deplete these nations' natural resources but also produced a wide range of environmental problems, including severe air, land, and water pollution; soil loss and degradation; the lowering of water tables and diversion of waterways; deforestation; habitat destruction; biodiversity loss; and the generation of enormous volumes of mine waste (Reed 1996; also see Battikha 2002). Indeed, the World Bank itself (Lele et al. 2000) made a similar argument regarding tropical deforestation in the 1990s, which it attributed, at least in part, to two important components of structural adjustment, trade liberalization and currency devaluation, which the Bank argued directly promoted timber exports and made it profitable to cut down forests to grow crops for export.

Because structural adjustment intentionally reduces government spending, it also limits the funds that governments are able to devote to managing and protecting natural resources and enforcing environmental regulations. This, of course, makes it difficult for adjusted nations to ensure that industrial activities, agriculture, and natural resource extraction are carried out in environmentally sustainable ways (Reed 1996; SAPRIN 2004; also see Battikha 2002). In addition, by increasing unemployment, lowering wages, and reducing government spending on social services, structural adjustment increases urban poverty. In the African nations included in the World Wildlife Fund's 1996 study, this resulted in the migration of many urban residents to rural areas, where they and an increasingly impoverished rural population overexploited local natural resources and environmental goods in order to survive (poor rural residents in the other nations included in the World Wildlife Fund's study were also forced to overexploit their local environments in order to survive).

Summarizing the Consequences of Structural Adjustment

The foregoing discussion clearly demonstrates that structural adjustment is strongly linked to a wide range of negative social, economic, and environmental outcomes in developing nations around the world. It would be foolish, of course, to think that these outcomes result solely from World Bank and IMF structural adjustment lending or to argue that factors internal to developing nations do not play a role in producing these outcomes. High levels of poverty, for example, existed in these countries long before the imposition of structural adjustment, as did natural resource extraction and agricultural export activities (though to a significant degree, these were the product of hundreds of years of colonial rule by the same Western nations that now control the World Bank and IMF). Elites in developing nations, like elites in wealthy nations, also have an interest in maintaining inequality; and in the context of global domination by Western nations and corporations, it is possible that developing nation elites view structural adjustment as one of the few sources of wealth available to them or, if they have some control over the flow of structural adjustment funds within their government and economy, as an important source of influence within these arenas.

Nevertheless, it is still the case that structural adjustment has fundamentally shaped the economies and policies of many developing nations, as evidenced not only by the degree to which many adjusted nations have been adjusted (see earlier discussion) but also by the incredibly large number of developing nations that have undergone structural adjustment. For instance, between 1980 and 1991, 89 developing nations entered into 566 structural adjustment loan agreements with the World Bank and IMF (Bello et al. 1999); and from 1988–2000, 33 of the world's 48 least developed countries borrowed money from the IMF under its SAF and ESAF programs, with 11 of these countries involved in SAF or ESAF programs for at least six of these years (UNCTAD 2000). Moreover, in 2000 some 70 countries were involved in IMF structural adjustment programs, with many of these nations having received IMF funding for at least 20 years (Meltzer 2000). It is thus fair to argue that through the mechanism of structural adjustment, the World Bank and IMF have played key roles in producing and intensifying a wide range of critically serious economic, social, and environmental problems in developing nations throughout the world.

In forcing developing nations to adopt structural adjustment policies crafted by foreigners, the World Bank and IMF have also impinged on the sovereign right of adjusted nations to develop economic, social, and environmental policies and economic, social, and government institutions as they see fit, while simultaneously undermining the rights of citizens in democratic adjusted nations to influence their nation's policies. As several researchers have noted, the large sums of money that some government agencies receive from the World Bank to implement structural adjustment reforms (Goldman 2005; G. Harrison 2004) and the fact that the World Bank and IMF deal almost exclusively with developing nations' central banks and ministries of finance (G. Harrison 2004; Woods 2006) also tend to distort debtor nations' internal political processes such that these processes tend to favor the interests of officials in these government departments and agencies.[13] The executive branches of some debtor nations are also able to "exact concessions from their legislatures" by arguing that legislators must adopt policies that fulfill their nation's structural adjustment loan obligations even though these legislators were not involved in World Bank and IMF loan negotiations and therefore never agreed to the conditions included in the loans (Meltzer 2000, 26). Thus, in addition to producing severe economic, social, and environmental harm in debtor nations, structural adjustment loans also impinge on these nations' sovereignty, distort these nations' political processes, and in democracies, weaken citizens' democratic rights.

Unfortunately, however, structural adjustment is not the only mechanism the World Bank and IMF use to shape social, economic, and environmental outcomes in developing nations. The World Bank, for example, also exerts influence on developing nations through loans that it makes for specific development projects and through its ability to constrain the production of knowledge regarding the economic, social, and environmental consequences of its projects. Therefore, in the next two sections of the chapter, I briefly examine these project-based loans and the methods the World Bank uses to both generate information and limit knowledge about its development projects.

Project-Based Loans and the Environment

In addition to providing developing nations with structural adjustment loans, the World Bank also provides funding to countries and corporations for specific economic development projects throughout much of the developing world. A large portion, though by no means all, of this project-based funding has been used to build roads and dams, generate electricity, extract oil, gas, and minerals, and harvest timber, though in many cases timber harvesting has resulted from World Bank loans for other activities (Bosshard et al. 2003; Rich 2009).[14] To illustrate the social, economic, and environmental consequences of World Bank economic development projects, this section of the chapter highlights several large development projects that the World Bank has funded over the past five decades.

The first two projects I discuss are the Polonoroeste and Carajas Grande projects in Brazil. In order to help finance the Polonoroeste project, located in the Brazilian state of Rondonia, the World Bank lent Brazil more than $443 million in the early 1980s, or roughly one-quarter of the project costs, to build a highway and access roads that would connect "Brazil's populous south central region with the rainforest wilderness in the northwest" (Rich 1994, 27). According to Jose Lutzenberger, a Brazilian agronomist and leading environmental activist at that time, the goal of the project was to resettle some of Brazil's 2.5 million landless rural poor to the Amazon to relieve social and political pressures in more populated areas of the country (House Committee on Science and Technology 1984). However, the more than 500,000 people who migrated to the Amazon in the 1980s as a result of the Polonoroeste project did not receive adequate relocation assistance from the government and were thus forced to clear large tracts of forest to grow crops that soon depleted the soil, leading them to cut down still more forest tracts. In addition, timber companies used the newly constructed highway and access roads to gain access to the region's forests (Keck 1998), such that between 1978 and 1991 the percentage of land in Rondonia that was deforested increased from 1.7% to 16.1% (Rich 1994), a statistic that is particularly troubling because deforestation is strongly associated with soil erosion, impaired water quality, biodiversity loss, and climate change.

The human costs of the Polonoroeste project were also severe. Influenza, measles, and tuberculosis swept through indigenous areas; 250,000 indigenous people and settlers contracted malaria; native peoples were displaced from their lands; and land conflicts and violence erupted between "rubber tappers, indigenous tribes, cattle ranchers, and colonists" (Rich 1994, 29).

The Carajas Grande project also produced severe human, social, and environmental harm. Carajas Grande was a regional development project built around an iron ore mine, a deep sea port, and a railroad connecting the two in Para State, Brazil. Because the goal of the project was not just to increase mineral exports but also to promote economic development in the larger region in which the mine, railroad, and port were located, the project also included the construction of two aluminum plants and the Tucurui dam, which provided electricity to the aluminum plants and the Carajas mine and which was originally designed to provide waterway transport of ore from the mine to the Vila do Conde seaport (Hall 1989).

In 1982 the World Bank provided $304.5 million in loans to Brazil to construct the mine, railroad, port, and several town sites for workers. The Bank also helped Brazil raise funds for the project from other official donors (Hall 1989; World Bank 1982). The project was completed in 1985, and within five years approximately 150,000 square kilometers of forest in the region had been cut down (Rich 1994).

One of the reasons the Carajas Grande project produced such severe deforestation was that upon completing the project, six pig iron plants were built to process ore from the mine. These plants consumed large quantities of fuelwood, which was drawn primarily from the local area. Deforestation also occurred because the project made the region accessible to timber companies, ranchers, and slash and burn agriculturalists. Cities that sprang up around the mine also created environmental problems. For instance, in the small city of Parauapebas, which was built to house mine company employees, "the surrounding hillsides [were] all but totally stripped of their tree-cover to supply building materials and fuelwood, giving rise to widespread soil erosion and, according to local inhabitants, changes in the micro-climate such as higher maximum temperatures and sparser rainfall" (Hall 1989, 168). Making matters worse, hundreds of thousands of small-scale gold miners flocked to the now accessible region. These miners destroyed streambeds and stream banks, decimated local fish populations, and polluted local waterways with silt and mercury, which in turn may have poisoned local indigenous people (Roberts and Thanos 2003, 152–153).[15]

The Tucurui dam, which as previously noted was a central feature of the Carajas Grande project, also harmed the environment in several important ways.[16] It diminished downstream water quality, prevented seasonal flooding, reduced downstream nutrient flows and fertilization processes, eliminated 19% of downstream fish species, flooded 2,850 square kilometers of land, and emitted significant volumes of greenhouse gases due to the existence of rotting vegetation in the dam reservoir (La Rovere and Mendes 2000).[17]

Dam construction also resulted in the resettlement of 25,000–35,000 people. These people were not adequately compensated for their losses and were relocated to areas with limited and inadequate survival options. Because dam construction increased the size of the local population, it also increased the incidence of alcoholism, sexually transmitted diseases, AIDS, malaria, schistosomiasis, and other diseases in the region. Increases in the mosquito population during the first several years of the dam's operation also contributed to the rise of malaria and other diseases (La Rovere and Mendes 2000).

The human and environmental effects of the Tucurui dam are not surprising given that, worldwide, dams are associated with much environmental degradation and human suffering. Research shows, for example, that dams negatively affect terrestrial ecosystems and biodiversity, that the reservoirs they create emit large volumes of greenhouse gases, and that they severely harm downstream aquatic ecosystems, due in part to their effect on river flow and water temperature. Dams also trap sediments and nutrients behind their walls, disrupt floodplain ecosystems, and often decimate fisheries that people rely on for food and income. Moreover, between 1950 and 2000, large dams physically displaced an estimated 40–80 million people around the world (28 million between 1986 and 1993), many of whom were not given new homes or adequate compensation for their losses (World Commission on Dams 2000).

As the world's largest and most prominent development institution, the World Bank has played a critically important role in constructing hundreds of dams around the globe, including some of the world's largest. In fact, between the end of World War II and 1993, "the Bank made 527 dam-related loans for a total of $58 billion (in constant 1993 dollars)," and between 1993 and 2002, "hydropower and irrigation accounted for 8% of all World Bank lending" (Bosshard et al. 2003, 27–28). It should not be surprising, then, that during the 1993 fiscal year, dams associated with active World Bank loans had displaced 1.23 million people, a number that does not include people displaced by dams funded by then inactive World Bank loans or by loans made after the 1993 fiscal year (World Bank 1994).[18]

Nor should it be surprising that the World Bank continues to fund socially and environmentally harmful development projects. For instance, between 1992 and 2004, the World Bank spent nearly $11 billion to help finance 128 oil, gas, and coal extraction projects that will likely produce enough fossil fuel to generate 37.45 billion tons of carbon dioxide emissions (Vallette et al. 2004); and between 1994 and January 2009, the World Bank provided or had plans to provide just over $5.3 billion in financing for the construction and expansion of 22 coal-fired electricity plants that were expected to produce more than 217 million tons of carbon dioxide every year, 77.8 million tons per year of which will represent entirely new emissions.[19] These emissions will, of course, contribute to climate change and are likely to increase the incidence of heart ailments, respiratory disease, and lung cancer in the areas around the plants. Indeed, the Environmental Defense Fund estimates that between 6,000 and 10,700 people are likely to die each year due to the combined emissions from these plants and 66 other coal-fired electricity plants financed by multilateral development banks and export credit agencies between 1994 and January 2009 (Penney et al. 2009).[20]

In addition to these projects, the World Bank has also played an important role in recent years in funding and organizing the Greater Mekong Subregion (GMS) Development Project (GMSDP), which "encompasses most of Cambodia and Laos, almost one-third of Thailand, and parts of China, Myanmar, and Vietnam" (Mekong River Commission 2011a). Long-term development plans for the GMS include the building of highways, bridges, and hydroelectric dams, the expansion of plantation agriculture, logging, and mining, and the fostering of trade, particularly in energy, between GMS countries. From 1992 to 2008, nearly $10 billion was invested in the GMSDP, almost $7 billion of which came from official donors such as the World Bank and the Asian Development Bank (ADB) and $6.4 billion of which (as of December 2005) went to just two sectors: transportation ($4.8 billion) and hydroelectric energy ($1.6 billion) (Soutar 2009).

Although the ADB has been the primary donor supporting the GMSDP, the World Bank has played a key funding and planning role in several important aspects of the project, including water resource management, trade and transport facilitation, labor migration, dam financing, and regional electrical power trade (Porter 2007; Ransley et al. 2008;

World Bank 2007). For example, the World Bank played an important role in securing financing for the Nam Theun 2 dam and hydroelectric power project in Laos (World Bank 2010);[21] and in anticipation of building dozens of other hydroelectric dams in Laos and opening up access to other natural resources in the country, the World Bank also provided extensive financing to Laos to "rewrite property rights laws, redesign state agencies, . . . redefine localized production practices, . . . [and] generat[e] new state authorities within [the Laotian government]" (Goldman 2005, 182–183). The World Bank also works actively with the ADB and the Mekong River Commission to design and implement regional development efforts (Mekong River Commission 2011b; Ransley et al. 2008; Soutar 2009).

The World Bank and ADB both claim that with the proper safeguards, hydropower and other development projects in the GMS can be implemented in environmentally sustainable ways that help alleviate poverty (Ransley et al. 2008; Stenhouse 2010; Vostroknutova 2010; World Bank 2007). As partial proof of this, the ADB reports that the percentage of people living on less than a dollar a day has declined dramatically in each of the six GMS countries since the inception of the GMSDP (Ransley et al. 2008). The World Bank and the ADB also describe the operational Nam Theun 2 dam as "a world's best practice dam" (Ransley et al. 2008, 32) that will "help protect the environment, . . . alleviate poverty in Laos, . . . and improve the social and environmental performance of the Lao hydropower sector" (International Rivers 2008, 4).

Evidence obtained from official and unofficial sources contradicts these claims. For instance, although it is true that the percentage of people living on a dollar a day has declined substantially in the region, the monetary gains that poor people in the GMS have experienced are in many cases more than offset by a decline in access to natural resources that they have also experienced as a result of dam-related flooding and the granting of private property rights in land and natural resources to commercial logging, agriculture, fishing, and hydropower interests (Cornford 2008; Ransley et al. 2008). The creation and granting of these previously nonexistent private property rights has also played an important role in depleting the natural resources that poor people in the region depend on for survival (Cornford and Matthews 2007). It is not surprising, then, that a study commissioned by the ADB found that economic conditions in Laotian villages had either remained the same or worsened between 2000 and 2006 and that in 2006 families were having greater trouble than previously obtaining food (Cornford 2008).

Development in the region has also produced or is projected to produce a wide range of other social, economic, and environmental problems. The Mekong River Commission estimates, for example, that the 11 dams currently planned for the lower Mekong will "turn 55 percent of the river into reservoirs, resulting in estimated agriculture losses of more than $500 million a year" (Petty 2011) and the forced relocation of approximately 88,000 people (International Rivers 2009). There are also concerns that damming the lower Mekong will affect the flow of vital nutrients through the river ecosystem, increase algae growth, stunt rice production, and prevent fish from accessing upstream spawning grounds. Damming the lower Mekong is also likely to devastate the region's fisheries, which on the one hand provide "between half and four-fifths of the animal protein consumed by the lower Mekong's 60 million residents" (International Rivers 2009, 2) and on the other hand contribute between $5.6 billion and $9.4 billion per year to the regional economy (Agence France-Presse 2011; Petty 2011).

One of the most controversial dams in the GMS is the Nam Theun 2, which the World Bank and the ADB both support and both designate as a "best practices dam." Nam Theun 2 began impounding river water in 2008 and commenced commercial hydroelectric operations in 2010, at which point it had flooded 450 square kilometers of land, including farmland, villages, and wildlife and wetlands habitat. Construction of the dam resulted in the forced relocation of 6,200 people to areas with poor soil and forest resources, caused an estimated 2,000 additional households to lose access to critical ecosystems resources, resulted in downstream erosion, flooding, and sedimentation, and increased wildlife and rosewood tree poaching in the formerly remote Nakai-Nam Theun National Protected Area (Lawrence 2008a, 2008b; McDowell et al. 2009; Ransley et al. 2008, 33). The easily anticipated flooding of downstream riverbank gardens along the Xe Bang Fai River[22] and the decline in fisheries and water quality that resulted from the building of the dam are also estimated to have negatively affected between 110,000 and 120,000 downstream villagers (International Rivers 2010; Ransley et al. 2008).

A supposedly independent panel of experts (McDowell et al. 2008) that has strongly supported the dam since 1997 reports that the standard of living of resettled villagers began to decline in 2007 and was likely to worsen in coming years. The panel also warned that degradation of downstream fisheries would likely result in swift and dramatic declines in protein consumption in 37 downstream villages and more gradual declines in villages along the Nam Theun River's tributaries (McDowell et al. 2008). And finally, because mitigation and compensation planning for affected villagers and ecosystems did not begin until 2008 (Ransley et al. 2008), three years after dam construction began and mere months before the dam reservoir began to fill, mitigation and compensation efforts did not commence until after anticipated declines in ecosystem productivity and protein consumption began to be felt (McDowell et al. 2008, 2009). Compensation efforts have also been underfunded and will likely be terminated in 2015 (International Rivers 2010) even though resettled villagers and those still living in their old homes will likely require continued assistance after that date.

Legitimizing the Nam Theun 2

An important question arises from the foregoing discussion. How is the World Bank able to designate Nam Theun 2 as a best practices dam and claim that it is concerned about poverty, human development, and the environment when the human, social, and environmental consequences of Nam Theun 2 and many other World Bank development projects are so dire? The answer, in short, is that (a) the World Bank is able to shape and control much of the information generated about its development projects and (b) unlike activist organizations that oppose these projects, the World Bank is widely viewed as an important and legitimate source of information and knowledge regarding economic development and the developing world. I address the first of these two points in the remainder of this section by summarizing the findings of a study conducted by Michael Goldman (2005) that examined the World Bank's ability to limit what the public knows about the Nam Theun 2 dam.[23] I address the second point later in the chapter.

Goldman begins by noting that when the Nam Theun 2 project was initially proposed, the World Bank hired an engineering firm that had previously worked with the largest private investor in the project to assess the feasibility and social and environmental consequences of the proposed dam. Due to pressure from nongovernmental organizations (NGOs) regarding the flawed nature of both this assessment and a subsequent assessment carried out by an engineering firm that had worked with the Bank in the past, the Bank eventually hired two NGOs, the International Union for Conservation of Nature (IUCN) and CARE International, to conduct new social and environmental impact assessments of the dam. The seemingly independent nature of these assessments was undermined, however, by the fact that IUCN and CARE International already supported Nam Theun 2, in large part because they believed that dam construction was inevitable and that failing to support the project might lead to their expulsion from Laos.

Moreover, while the circumstances within which IUCN and CARE International were operating suggest quite strongly that neither of them was in a position to independently assess the project, IUCN experienced an even more serious conflict of interest in its role as independent assessor because it stood to receive tens of millions of dollars from the sale of electricity generated by the dam if the dam were constructed, with IUCN planning to use this money to "design and run a series of [large] National Biodiversity Conservation Areas" (Goldman 2005, 165) that it had good reason to believe would not be developed without revenue from the dam.

Further undermining the independence of these and other subsequent assessments, the terms of reference under which impact assessors (the research consultants who were paid to conduct the actual assessment research) were hired by the NGOs, the World Bank, and the Laotian government tended to be highly restrictive, echoing a problem that Goldman argues is fairly common at projects supported by the World Bank:

In exchange for high salaries, unique research opportunities, and access to formerly inaccessible research sites, [impact assessors are told] exactly what kinds of information are needed, a time frame for completing the research (and by implication, how long the researcher can be in the field), and a deadline for the written report. Ownership and circulation are also important dimensions of the [terms of reference]: the direct contractor—be it the Bank, the borrowing government, an engineering firm, or an NGO—is given exclusive right of ownership over the product as well as the raw data. Legally, [impact assessors] cannot use the data for research or distribute the findings without permission from the contractor. (Goldman 2005, 160)

The result, Goldman demonstrates, is that in Laos impact assessors generally lacked sufficient time to thoroughly or adequately study the potential negative consequences of the Nam Theun 2 dam, and when they did uncover social or environmental consequences that might jeopardize the project, these consequences often fell outside the boundaries of what they were supposed to study (outside the terms of reference) and thus were not included in the reports they wrote.

This rather subtle practice of information suppression was bolstered in at least one instance, Goldman notes, by an IUCN decision to simply not circulate a contractor report that found that resettled villagers would no longer be able to supplement their rice production with food from foraging, hunting, and fishing and, thus, might not survive resettlement. To replace this censored report, the dam investors (rather than IUCN) hired another consultant, who concluded that resettlement would not adversely affect the local population.

Goldman further notes that because the World Bank is only interested in specific kinds of information and specific types of development indicators, the knowledge generation structures it creates tend to ignore or dismiss the value and importance of noncommodified social and economic relationships, which not only are greatly harmed by most World Bank development projects but are also the types of relationships that communities affected by these projects often value the most. The World Bank's knowledge generation structures, which prioritize scientific and market rationality, also tend to define noncommodified relationships and activities as being environmentally harmful. For example, Goldman notes that in the Greater Mekong region the subsistence activities of local groups such as forest dwellers are often described by the Bank as being environmentally degrading, while environmentally devastating activities such as export-oriented rice cultivation, capital intensive electricity generation, and timber harvesting are generally described as being environmentally benign or sustainable. Finally, Goldman finds that the type of information generation and suppression activities found at Nam Theun 2 are not unique to that project but are instead found at many other World Bank development projects around the world. It is thus clear that the World Bank regularly uses information generation and suppression to help it frame its projects and activities in positive social and environmental terms.

The evidence presented in this and the preceding sections thus demonstrates not only that World Bank development projects severely harm individuals, communities, and the environment throughout much of the developing world but also that the World Bank, through its knowledge generation and suppression activities and its relationships to NGOs, is able to limit the creation and dissemination of information that calls into question the social and environmental desirability of its projects. The World Bank's ability to garner support from NGOs that are supposedly devoted to helping poor people and saving the environment also provides the World Bank with these organizations' stamp of approval, an important form of legitimacy for an organization that has been severely criticized for decades for its widespread human rights and environmental abuses. The result of all this is that the World Bank is able to frame its development projects as being socially and environmentally sustainable, even "best practices," projects, and thus good for the environment and the poor even when these projects severely harm poor people, poor communities, and the environment.

Who Controls the World Bank and IMF?

In the preceding sections of this chapter, I demonstrated that the World Bank and IMF, through their lending and knowledge generation and suppression policies, play key roles in producing social, economic, and environmental harm in much of the developing world. I further demonstrated that World Bank and IMF policies undermine the sovereignty of debtor nations, distort their political processes, and in democratic debtor nations, weaken citizens' democratic rights. The question still remains, however, of whether the World Bank and IMF are undemocratic and controlled by elites and, if so, who these elites are and how they exert disproportionate influence over these two institutions. I answer this question in two stages. I first show that the U.S. and a handful of other wealthy nations use a set of organizational mechanisms, based in the voting and funding structures of the World Bank and IMF, to undemocratically control much of what the World Bank and IMF do, with the U.S. exerting greater influence over these two institutions than does any other nation in the world. I then shift my focus to the U.S. and ask which groups in the U.S. played the greatest role in shaping U.S. support for the neoliberal structural adjustment policies that the U.S. successfully pushed the World Bank and IMF to pursue in the 1980s and early 1990s.

As we shall see, the major force behind these neoliberal policies was a set of conservative think tanks in the U.S. that provided their corporate elite funders with the ideological arguments and practical political strategies that these elites needed to achieve their economic and political goals. We will also see that those U.S. government officials most directly involved in shaping World Bank and IMF policy in the 1980s were members of the power elite who had strong ties to the financial sector of the U.S. economy, thereby providing the power elite and the U.S. financial sector with direct, disproportionate, and undemocratic influence over the United States' World Bank and IMF policy. I thus conclude that conservative think tanks and direct ties to the U.S. government were two critically important organizational and network-based mechanisms used by economic elites in the United States to develop and convey to the U.S. the socially, environmentally, and politically harmful policies that the U.S., using its organizational advantages at the World Bank and IMF, then forced these two institutions to implement.

U.S. Influence over the IMF and World Bank

U.S. influence over the IMF and World Bank is based in the organizational, voting, and funding structures of these two institutions and in the power the U.S. has in the international arena, all of which I discuss in turn. In terms of organizational structure, the World Bank has five arms: the International Bank for Reconstruction and Development (IBRD), the International Development Association (IDA), the International Finance Corporation (IFC), the Multilateral Investment Guarantee Agency (MIGA), and the International Centre for Settlement of Investment Disputes (ICSID).[24] The World Bank is headed by a president and the IMF by a managing director, each of whom is formally appointed by her or his institution's board of directors. However, by tacit agreement the World Bank president is selected by the U.S. and the IMF managing director by the nations of western Europe. In addition, the IMF's second highest ranking officer, the first deputy managing director, is usually a U.S. citizen.

The IMF and World Bank each have a board of governors whose membership includes a minister of finance, a minister of development, or a central bank governor from each member nation. These boards each meet once a year and oversee the general operations of their respective institutions.[25] However, much of the authority vested in these boards is delegated to a board of directors at each institution that oversees its respective institution's daily operations.[26] For example, among other responsibilities, the IMF board of directors helps set IMF strategy, conducts surveillance of member countries' policies and economies, conducts institutional oversight activities, and approves the use of IMF resources, including the granting of loans negotiated by IMF staff (Independent Evaluation Office of the IMF 2007; Rastello 2010). The board of directors at the World Bank is responsible for similar tasks, including considering and approving "IBRD loans and guarantees, IDA credits and grants, IFC investments, MIGA guarantees, and policies that impact the World Bank's general operations" (World Bank 2011a).

The board of directors at each institution had 24 members for much of the period covered in this chapter,[27] with 16 of the directors at each institution representing multiple nations and the remaining directors each representing a single nation (the eight nations that each had their own board member as of October 2010 were the U.S., the UK, Japan, France, Germany, Saudi Arabia, China, and Russia). The voting power of each executive director equals the sum of the voting power of the nations he or she represents on the board, with the voting power of each IMF and World Bank member nation roughly proportional to its global economic power. Thus, in 2010 the U.S. and its IMF executive director controlled 16.17% of the IMF board's votes, the executive directors of the UK, France, Germany, and Japan each controlled between 4.7% and 5.8% of the board's votes, and the executive directors of Russia, Saudi Arabia, and China each controlled between 2.6% and 3.6% of the board's votes. The remaining directors each represented multiple countries, with four directors, representing a total of 38 countries, each controlling between

4% and 5.01% of the board's votes; 11 directors, representing a total of 116 countries, each controlling between 2% and 4% of the board's votes; and one director, representing 22 African nations, controlling 1.62% of the board's vote.

Voting power at the World Bank is similarly biased, with the U.S. (in 2006) controlling 16.4% of the votes on the IBRD's board of directors and the UK, France, Germany, and Japan each controlling between 4.3% and 7.9% of the board's votes. Of the remaining directors, three controlled between 0.28% and 2% of the votes, four controlled between 2% and 3% of the votes, seven controlled between 3% and 4% of the votes, and five controlled between 4% and 5.04% of the votes, with most of these directors representing multiple countries.

Not only is voting power at the IMF and World Bank highly skewed, but many of these institutions' most important board decisions require an 85% majority, which makes the U.S. the only nation in the world that has unilateral veto power over these decisions. Moreover, high income nations currently hold over 60% of the votes on the executive boards of the World Bank's IBRD, IDA, and IFC, with middle income countries holding roughly one-third of the votes on each of these boards and low income countries holding between 3% and 11.3% of the votes on these boards (Horton 2010). The situation is only slightly less skewed at the IMF, where high income nations now control just over 55% of the votes (IMF 2011). Thus, a small group of wealthy nation executive directors holds enough votes to approve or veto World Bank and IMF board decisions even in situations where a 50% majority is needed to do so.[28]

Some people argue that this skewed voting structure is not problematic or undemocratic because the boards of directors at the World Bank and IMF operate by consensus. However, this is a highly misleading argument. First, as I discuss in more detail later in the chapter, it is likely that developing nations are quite reluctant to stake out positions at the World Bank and IMF that run counter to those held by the U.S. and other powerful nations because these powerful nations hold the key to World Bank, IMF, and other resources that developing nations need. Second, over the past 12–15 years, the IMF board has moved to a decision making process in which the chair of the board, who is either a European or U.S. official, subjectively determines or interprets the opinion of the board by informally polling its members (Buira 2003; Chowla

2007), giving much more weight than did consensus decision making to those executive directors who control the most votes. Third, much of the decision making power at the IMF lies outside the formal structure of the IMF and with the finance ministers of the wealthy Group of Seven (G-7) nations: the U.S., the UK, France, Germany, Italy, Japan, and Canada. Ngaire Woods notes, for instance, that "a subgrouping of G-7 Finance Ministry deputies regularly convenes to discuss the issues confronting the [IMF] and the world economy, updated and advised by the U.S.-appointed first deputy managing director of the IMF. It is this group rather than the formal oversight body—the International Monetary and Financial Committee (IMFC)—which guides the institution, or as a report in 2004 puts it, assumes the strategic guiding role in respect of the IMF" (2006, 191). Thus, at the IMF the G-7 nations provide overall strategic guidance and control 41% of the votes on the board of directors, while the U.S., on its own, can block many of the most important items or proposals brought before the board.

The situation is not much different at the World Bank, where executive directors usually rubber stamp matters that staff and management place before them. There are three main reasons for the board's acquiescence on most matters. First, developing nation executive directors generally support the granting of loans and other forms of financial assistance to nations they do not represent so that other directors will do the same for them. Second, it is very difficult for most developing nation directors, who generally control only 2%–4% of the votes and represent multiple countries, to coordinate their activities with other similarly "disadvantaged" directors to achieve the 15% voting bloc they need to veto board proposals (it is impossible for them to achieve a 50% voting bloc on their own). It is similarly difficult for nations that share a single director to coordinate their interests so that they can present a set of mutually agreed on concerns and policies to the board. The existence of these two disadvantages (the difficulties of coordinating interests and activities *within* and *across* developing nation directorships) means that it is very difficult for all but the most powerful countries and directors at the World Bank (and IMF) to accomplish their goals. Indeed, Woods argues that the situation is so bad for most World Bank and IFM executive directors that there is very little incentive for them "to do their job thoroughly" (2006, 191).

Third, it is likely that when considering what policy positions to take at the World Bank and IMF, developing nation executive directors are deeply concerned about the United States' ability to punish the nations these directors represent, both within the World Bank and IMF and through other channels. For example, the informal polling that replaced consensus decision making on the IMF executive board forces developing nation directors to specifically state their policy positions (Vreeland 2007), which they may be reluctant to do given U.S. power. In addition, research demonstrates that at the WTO, developing nations often accede to the wishes of the United States because the United States is willing to punish developing nations that do not support its WTO policies (Kwa 2002; Narlikar 2001), and there is no reason for developing nations to expect that the U.S. is unwilling to do the same at the World Bank and IMF. Research also shows that developing nations that are allied with the U.S. or that support U.S. policy at the UN tend to receive more favorable treatment from the World Bank and IMF than do other developing nations (Dreher and Jensen 2007; Kilby 2009). It is thus reasonable to expect that developing nations that support U.S. policy at the World Bank and IMF also receive more favorable treatment at these institutions.

The result of all this is that most matters brought to the attention of the Bank's board of directors have already been decided informally before being brought to the board, with the U.S. and other wealthy nations much more likely than middle- and low-income nations to be included in the informal negotiations that precede board meetings. Thus, as Woods notes of both the IMF and the World Bank, "A loan that [does] not meet with US approval would seldom be presented to a Board for discussion. Before getting that far, in most (but not all) cases staff and management would have been in dialogue with those whose agreement was necessary for the loan to go through" (2001, 87).

Consistent with this argument, research demonstrates that whether or not a developing nation receives an IMF or World Bank loan (Bird and Rowlands 2001; Dreher and Sturm 2006; Dreher et al. 2006; Fleck and Kilby 2006; Harrigan et al. 2006; Killick 1995; Stiles 1991), the size of the loan the nation receives (Andersen et al. 2006), the number of conditions attached to the loan (Dreher and Jensen 2007), and the degree to which loan conditions are enforced by the IMF and World Bank (Kilby

2009; Stone 2002, 2004; Vreeland 2005) depend in large part on whether the nation is a U.S. ally, whether it supports the U.S. position on United Nations votes that the U.S. publicly states are important to it, whether it receives U.S. foreign aid, and whether the U.S. or another wealthy nation intervened on its behalf during the loan approval process.[29] Indeed, in the 1980s, when the IMF and World Bank adopted structural adjustment as their main tool for addressing the developing world's debt crisis, the U.S. possessed so much power at the World Bank that "any signal of displeasure by the U.S. executive director ha[d] an almost palpable impact on the Bank leadership and staff, whether the signal [was] an explicit complaint or simply the executive director's request for information on a problem" (Ascher 1992, 124, quoted in Woods 2000, 134).

The power that the U.S. and other wealthy nations hold at the World Bank and IMF does not derive solely from their voting power, however. It also derives from the fact that the World Bank and IMF rely on wealthy nations for funding.[30] The IMF, for example, is dependent on "quota" contributions that member nations make to the organization for much of its budget, with wealthier nations making the largest of these contributions. Member nations' quota contributions, which determine each nation's voting power in the IMF, are reviewed every five years and can be increased only with an 85% majority vote of the executive board. This, of course, gives the U.S. significant influence over the IMF since the U.S. is the only nation in the world that can veto this request on its own and thus deny the IMF the funds it believes it needs to accomplish its mission.

The World Bank is less dependent than the IMF on funding from member nations, with much of the money it uses for research, administrative expenses, and loans coming from world capital markets, retained earnings, and repayment of previously disbursed loans. However, the World Bank's International Development Association relies entirely on funding from wealthy nations, and over the years the U.S. has influenced the policies of the IDA and the rest of the World Bank by threatening to withhold or reduce its IDA contributions during IDA replenishment negotiations, a threat that is especially potent because any decline in U.S. contributions can, by agreement, be followed by proportionally equivalent declines in IDA funding from other contributing nations (see Bello et al. 1999, Woods 2006, and the following sections for examples of this).

It is thus apparent that the organizational, voting, and funding structures of the World Bank and IMF provide the U.S., and to a lesser degree other wealthy nations, with the power to strongly shape World Bank and IMF lending decisions, including, as we shall see in the following sections, their respective decisions to adopt structural adjustment as their key policy response to the debt crisis sweeping the developing world in the 1980s. The U.S., of course, continues to wield enormous power at the World Bank and IMF, which helps explain these institutions' continued reliance on structural adjustment. But rather than examining U.S. influence over structural adjustment from the 1980s to the present, the following sections focus solely on the 1980s, when structural adjustment lending was first used to incorporate developing nations into the global neoliberal order championed by the United States and economic elites within the United States.

U.S. Influence over World Bank and IMF Structural Adjustment Policy in the 1980s

In order to best explain the role the U.S. played in shaping World Bank and IMF structural adjustment policy in the 1980s, I first describe the general policy positions the U.S. took during that decade vis-à-vis the World Bank and the developing world. I then demonstrate that virtually all the key decisions that led the World Bank and IMF to adopt structural adjustment as their key policy response to the debt crisis were made by the U.S., with the IMF playing a supporting role in the decision making process and the World Bank playing virtually no role at all.

However, before doing these things, it is important to point out that because the U.S. played a more central role than any other actor in shaping World Bank and IMF structural adjustment policy in the 1980s, and thus in shaping the basic structure of much IMF and World Bank lending since that time, the U.S. bears the lion's share of the blame for the devastating consequences these loans have had on much of the developing world.[31] It is also important to point out that in addition to being used to address the developing world's debt crisis, structural adjustment was also used by the U.S. to exert control over the governments and economies of the developing world.

U.S. POLICY VIS-À-VIS THE WORLD BANK AND THE
DEVELOPING WORLD

The late 1970s and early 1980s saw the rise of conservative governments in several of the West's leading nations, with Ronald Reagan coming to power in the U.S. in 1981, Margaret Thatcher coming to power in the UK in 1979, and Helmut Kohl becoming chancellor of Germany in 1982. To various degrees, these leaders and their conservative governments all supported neoliberal policies designed to drastically reduce government intervention in the market. However, the Reagan administration was particularly concerned by the actions of developing nation governments, which it believed were intervening much too strongly in their markets, thereby making it difficult for multinational corporations to invest and do business in these markets. The administration was also unhappy with World Bank lending policies that, in its view, did not place enough pressure on developing nations to open up their markets to corporations and foreign investors (Bello et al. 1999).

Thus, in 1982, as part of its efforts to reduce government intervention in developing nation markets, open up the economies of the developing world to U.S. corporations, and force the World Bank to change its lending policies, the U.S. drastically cut its financial contribution to the World Bank's IDA, which provides loans and grants to the world's poorest nations (Bello et al. 1999). This decision resulted in an overall reduction in IDA funding of between $1 billion and $1.5 billion as other nations followed the United States' lead. And to further signal its displeasure with the World Bank, the Reagan administration cut U.S. IDA contributions again a few years later (Kapur et al. 1997).

The World Bank also experienced an important change in leadership in 1981 when its longtime president, Robert McNamara, was replaced by former Bank of America president and chief executive officer Alden W. Clausen. Clausen was a stronger believer than McNamara in limiting government intervention in the market, and consistent with this belief, he hired Anne Krueger as World Bank chief economist and vice president for research. A "distinguished representative of the neoclassical school of development and trade policy, [Kreuger placed] a heavy pro-markets, anti-[government] . . . imprint on the Bank's research and policy-analysis programs" (Kapur et al. 1997, 511).

Thus, in the early 1980s two important events occurred at the World Bank. First, it experienced a sea change in ideology as McNamara and his staff were replaced by Clausen, Krueger, and their staffs. Second, despite the relatively close ideological alignment that now existed between the World Bank leadership and the Reagan administration, the administration still sent strong signals to the World Bank and its new president that the U.S. had no qualms about punishing the Bank if the U.S. was unhappy with it. To make sure the Bank understood this lesson, the Reagan administration sent the same signal to Clausen's successor at the Bank, Barber Conable, when he became World Bank president in 1986. In this case, however, the U.S. delivered its message by leading the Bank's board of directors in its refusal to approve the Bank's administrative budget for the following year.

Further ensuring that the World Bank and other multilateral development banks (MDBs) understood the U.S. position on MDB lending policy and neoliberal policy reform, a 1982 U.S. Treasury Department report, summarized by Catherine Gwin, noted,

> U.S. support for the MDBs should be designed to foster greater [developing nation] adherence to open markets and greater emphasis on the private sector as the main vehicle for [economic] growth. . . . The United States should [also] work to ensure that loan allocations [to developing nations are] made conditional on policy reforms in recipient countries. Finally, the United States should reduce its expenditures on the banks. [Specifically], the United States should develop a plan to reduce and eventually phase out new paid-in capital for the hard loan windows of all MDBs. And it should reduce, in real terms, its future participation in all MDB soft loan windows, especially the IDA. (Gwin 1997, 230–231)

Given the Reagan administration's neoliberal economic philosophy, its desire to weaken developing nation governments, and the strong pressure it brought to bear on the World Bank, it should come as no surprise that "by 1986 all new World Bank loans were essentially policy oriented [loans]" (Kolko 1988, 272) or that the authors of the definitive historical account of the World Bank note that by the mid- to late 1980s "adjustment loans [had become] the Bank's program loans of choice" (Kapur et al. 1997, 518). Nor should we be surprised to learn from Stanley

Fischer, Anne Krueger's replacement as World Bank chief economist, that "the US squelched [World Bank] research on [developing nation debt relief] during the mid-80s. ... [The World Bank] had to keep [this] research quiet, because the institution was under political orders (not only from the US, also the Germans, and the Brits) not to raise issues of debt relief" (Kapur et al. 1997, 1195).

Thus, by the mid-1980s the main lending policy adopted by the World Bank (and IMF) achieved two of the Reagan administration's chief foreign policy goals, reducing developing nation intervention in the market and opening up developing nations to foreign trade and investment. At the same time, the World Bank could not openly discuss debt relief, a policy option that might have allowed developing nations to avoid structural adjustment, and developing nations could not receive World Bank, IMF, and many other forms of assistance without first agreeing to undergo structural adjustment.

Nevertheless, the U.S. probably would not have been as successful as it was in getting the World Bank and IMF to pursue structural adjustment if not for the Latin American debt crisis that erupted in 1982. This crisis came to a head in August of that year when Mexico declared to the U.S. and IMF that it could no longer pay back its loans to commercial and official creditors, thereby threatening the stability not only of several major U.S. and Japanese banks but of the international financial system as a whole.

ADDRESSING THE DEBT CRISIS

The grave threat posed by Mexico's imminent loan defaults resulted in a flurry of negotiations between the U.S. government, the IMF, the Bank for International Settlements (BIS), the G-10 central bank governors,[32] Mexico, and the world's major commercial banks. However, as several histories make clear (most importantly Boughton 2001 and Kapur et al. 1997), the lead actors in these negotiations were U.S. Federal Reserve Chairman Paul Volcker, U.S. Secretary of the Treasury Donald Regan (and his senior staff), IMF Managing Director Jacques de Larosiere, and Mexican Secretary of Finance and Public Credit Silva Herzog, with Volcker taking the lead role in developing a solution for the crisis and Volcker and De Larosiere taking the lead in organizing negotiations between the key players.

Defining the Mexican debt crisis as a short-term crisis in liquidity, the U.S. and IMF determined that short-term IMF loans (through the IMF's Extended Fund Facility) tied to strict structural adjustment conditions needed to be a key component of any solution to the crisis. Thus, for Mexico to receive the funds it needed to avoid defaulting on its loans, it would have to "reduce public sector expenditure and investment, eliminate government subsidies, increase the cost of goods supplied by the government, increase income and sales tax, set positive real interest rates to discourage capital flight and increase savings, rationalize and stabilize the exchange rate, and reduce inflation" (Woods 2006, 48).

Because this basic approach was adopted by the IMF in its dealings with all developing nations experiencing severe debt problems in the early 1980s (Boughton 2001; Gold 1988), it led to a great expansion in IMF structural adjustment lending at that time. However, the World Bank did not become a major player in the attempt to resolve the Latin American debt crisis until the mid-1980s when it became clear that the initial solution to the crisis was not working and that a new solution was needed.

Once again the solution for the crisis was developed primarily by the U.S. (between the spring and fall of 1985), with the U.S. Treasury and Federal Reserve working together to develop an ambitious plan that was then partially scaled back (to ensure support from other powerful nations) in consultation with IMF Managing Director Jacques de Larosiere (Boughton 2001). This scaled back plan was presented to U.S. bankers by Federal Reserve Chairman Paul Volker and U.S. Secretary of the Treasury James Baker on October 1, 1985, with Baker presenting the plan to the IMF's Interim Committee five days later.[33] The Baker plan, as the solution was called, was approved, with only minor caveats, by the IMF's executive board on November 13, 1985, only 12 days after the board first saw a summary of the plan and the first time it officially discussed the plan; and it received public support from the heads of the IMF and World Bank just a few weeks later, on December 2, 1985 (Boughton 2001; Rieffel 2003; Woods 2006).

Although comprising several related policies, what is most important for this discussion is that the Baker plan called for increased overall levels of structural and sectoral adjustment lending. It also called on the World Bank and Inter-American Development Bank to take lead roles in disbursing these new structural and sectoral adjustment loans (Boughton 2001; Rieffel 2003; Woods 2006), thus drastically increasing the World Bank's structural adjustment portfolio.[34]

Once again, however, the U.S. plan did not solve the debt crisis. Thus, in March 1989 U.S. Secretary of the Treasury Nicholas Brady proposed yet a third plan, the Brady plan, to address the crisis. This plan was developed almost entirely by U.S. officials, under the direction of Brady's deputy, David C. Mulford, in the period between George H. W. Bush's November 1988 presidential election victory and his inauguration in January 1989 (Boughton 2001; Kapur et al. 1997). As was the case in 1985, the U.S. consulted with the IMF's managing director (Michel Camdessus) as well as with IMF staff before unveiling its plan so as to make it more acceptable to all the important parties (Boughton 2001). It is unclear, however, to what degree Mulford or Brady consulted with the World Bank prior to announcing the plan. Nevertheless, the definitive history of the World Bank notes that by 1988 "the Bank's management [had become] unwilling to take any steps that might jeopardize its brittle relationship with its major shareholders, especially [the U.S.]" (Kapur et al. 1997, 648). By 1988 the Bank had also decided to take a reactive rather than proactive stance in its handling of the debt crisis, and its chief economist had recently asked its president whether he should keep Bank research on debt relief secret, presumably to avoid angering the U.S., the UK, and Germany (Kapur et al. 1997). It thus appears that as in 1982 and 1985 (see note 34), the U.S. did not have to consult with the Bank in any serious way before unveiling its plan.

Brady publicly presented his plan for the first time on March 10, 1989, and the following week the IMF's executive board met to discuss it. Several critical objections to the plan were raised by the board, which were worked out two weeks later at a "regularly scheduled meeting of the ministers of finance and central bank governors of the G-7 countries, . . . making [the plan's] acceptance the next day by the [IMF's] Interim Committee all but inevitable" (Boughton 2001, 494; see note 33 for a description of the Interim Committee).[35] Upon receiving the Interim Committee's endorsement, the plan went back to the IMF's board of directors, where it was "adopted essentially intact" (Boughton 2001, 498). The plan also received support from the World Bank, though the Bank was somewhat dubious about the plan's viability (Kaput et al. 1997).

As was true of the Volcker and Baker plans, the Brady plan relied heavily on structural adjustment. But unlike prior U.S. plans, it also called for partial debt relief for at least some debtor nations. Debt relief had been recognized for some time as an important component of any feasible solution to the crisis by at least some wealthy nations and World Bank economists. However, it could not be included in any politically viable solution to the crisis until the U.S. gave its approval for such a strategy, which it did in 1989. Thus, even though debt relief was not initially proposed by the U.S., the U.S. played a lead role in determining when debt relief could become part of World Bank and IMF policy by exerting or withholding its veto power over any proposals that included debt relief as a solution to the crisis (Boughton 2001; Woods 2006).[36]

It is thus clear that in the 1980s the U.S. played a more important role than any other actor in both determining how the World Bank and IMF addressed the developing world's debt crisis and ensuring that structural adjustment played a central role in any solution to the crisis. In fact, U.S. influence over the World Bank was so great during the 1980s that Paul Volker claims the U.S. Federal Reserve and Treasury Department essentially "directed the lending of the bank" at that time (Gwin 1997, 236). Whether or not this is true and despite the fact that the U.S. had to negotiate with commercial bankers, the IMF, and other powerful nations to achieve its goals, it is nevertheless the case that at all three key decision making moments in the debt crisis (1982, 1985, and 1989), the U.S. played the key role in determining how to solve the crisis, either by proposing policies that were eventually adopted by the IMF and World Bank or by allowing previously proposed ideas to become World Bank and IMF policy.

The question still remains, however, of whether in the United States, structural adjustment was solely the policy goal of political elites or whether economic elites also played an important role in promoting it. It is to this question that I turn in the following section.

Economic Elites, Neoliberalism, and U.S. Structural Adjustment Policy

To help determine what, if any, role economic elites in the U.S. played in promoting structural adjustment, I turn to power structure theory, which I described in some detail at the end of chapter 2. Power structure theory holds that a small group of politically active economic elites, known as the power elite, disproportionately influence U.S. policy making both by taking an active role in U.S. decision making (as U.S. government officials and presidential advisers) and by providing funding for a set of influential policy planning organizations that connect the power elite to the U.S. government and provide them with the expertise they need to identify and achieve their policy goals. Members of the power elite include individuals who serve as directors on multiple corporate boards of this nation's most prominent corporations, banks, and law firms or who serve on a prominent corporate board while being active and influential in directing the operations of some segment of the policy planning network (Domhoff 2002).

If power structure theory is correct, the power elite should have influenced U.S. debt crisis policy in two key ways. First, members of the power elite should have taken a direct and active role in shaping this policy by being placed in key positions of decision making authority within the Reagan and Bush administrations. Second, they should have played an indirect, but critical, role in shaping this policy by funding think tanks and other policy planning network organizations that developed neoliberal structural adjustment policies that were then conveyed to and adopted by the White House. As I will demonstrate, the power elite involved themselves in debt crisis policy making in both of these ways.

To demonstrate this, I first examine the elite backgrounds of the United States' key debt crisis decision makers: Federal Reserve Chairman Paul Volcker and U.S. Treasury Secretaries Donald Regan, James Baker, and Nicholas Brady.[37] Paul Volcker was clearly a member of the power elite when he was nominated to head the Federal Reserve in 1979. At that time he was a member of the Trilateral Commission, on the board of directors of the Council on Foreign Relations, and on the board of trustees of the Rockefeller Foundation, three of the most prominent policy planning organizations in the U.S. in the 1970s. He was president of the Federal Reserve Bank of New York from 1975 to 1979, was the under secretary of the U.S. Treasury from 1969 to 1974, and had been the vice president and director of forward planning at Chase Manhattan Bank from 1965 to 1969 (Committee on Banking, Housing, and Urban Affairs 1979).

Donald Regan was also a member of the power elite when President Reagan appointed him to head the Treasury Department in 1981. At the time of his appointment, Regan was chairman and CEO of Merrill Lynch & Company and was affiliated with three of the most prominent policy planning organizations in the country: he was a member of the Council on Foreign Relations, a trustee of the Committee for Economic Development, and a member of the Policy Committee of the Business Roundtable, which determines what issues the Roundtable will address in its efforts to shape U.S. policy. In addition, he was vice chairman of the board of directors of the New York Stock Exchange from 1972 to 1975 and chairman of the trustees of the University of Pennsylvania from 1974 to 1978.

Nicholas Brady was likewise a member of the power elite when he was nominated to become secretary of the Treasury in 1988. At that time he was chairman of the board and managing director of Dillon, Reed & Company, a prominent Wall Street investment banking firm. He was also chairman of Purolator Courier Corporation and sat on the boards of directors of Doubleday & Company and NCR Corporation, NL Industries Inc. Prior to his appointment in 1988, Brady served on several presidential commissions, including the President's Commission on Strategic Forces (1983), the National Bipartisan Commission on Central America (1983), the Commission on Security and Economic Assistance (1983), and the Blue Ribbon Commission on Defense Management (1985). In the early 1980s he also served on the boards of Rockefeller University, Bessemer Securities Corporation, and the Economic Club of New York. He later became a member of the Council on Foreign Relations.[38]

James Baker's profile is somewhat different from those of Volcker, Brady, and Regan. Although clearly a member of the power elite in later years and a central player in U.S. politics and policy making at the time of his appointment to head the U.S. Treasury, Baker was not a member of the power elite in the mid-1980s. Nevertheless, consistent with power structure theory, three of the United States' four key debt crisis decision makers in the 1980s—Volcker, Brady, and Regan—were members of the power elite. In addition, prior to entering government service, Volcker, Brady, and Regan each ran or helped run a large U.S. bank or Wall Street investment firm, suggesting that the U.S. financial sector had key people in the Reagan and Bush administrations looking out for their debt crisis interests.

In addition to directly determining how the U.S., the World Bank, and the IMF would address the Latin American debt crisis, members of the power elite also indirectly influenced World Bank and IMF policy by funding a set of moderate conservative and ultraconservative policy planning organizations that played an important role in influencing U.S. policy in the 1970s and 1980s (see chapter 2 for a discussion of how policy planning organizations fit into power structure theory).

Prior to 1980 the most influential think tanks and policy planning groups in the country were moderate conservative organizations such as the Council on Foreign Relations, the Brookings Institution, the Trilateral Commission, and the Committee for Economic Development (Domhoff 2002). These moderate conservative policy planning organizations (with the exception of the Trilateral Commission) remained influential throughout the 1980s. However, in the late 1970s and early 1980s their influence declined relative to that of ultraconservative policy planning organizations as corporate leaders became increasingly alarmed by declining corporate profits, high business taxes, the strength of labor, government interference in corporate affairs, and a political system that they felt was giving too much power to nonbusiness interests (Ferguson and Rogers 1986; Himmelstein 1990; Peschek 1987).

Corporate leaders responded to this situation by mobilizing themselves behind an essentially neoliberal economic agenda, which they promoted in a variety of ways, the most important of which was to dramatically increase their funding of ultraconservative policy planning organizations (Himmelstein 1990). The most important of these policy planning organizations in the late 1970s and early 1980s were the American Enterprise Institute (AEI), the Hoover Institution, and the Heritage Foundation, all of which espoused a radical neoliberal agenda that sought to promote individual freedom and economic growth by drastically reducing government intervention in the economy (Himmelstein 1990; Peschek 1987).

Thus, between 1970 and 1983 the Hoover Institution's annual budget increased from $1.9 million to $8.4 million, while AEI's budget increased from $0.9 million to $10.6 million. The Heritage Foundation, which was not founded until 1973, also had a $10.6 million budget in 1983 (Himmelstein 1990). And unlike in earlier decades, when funding for ultraconservative policy planning organizations came from a relatively small group of corporate elites, the funding and leadership for these organizations now came from a broad spectrum of the most powerful corporations, foundations, and CEOs in the country, including the Ford Motor

Company, Hewlett-Packard, Chase Manhattan Bank, Standard Oil of California, Dow Chemical, Mobil Corporation, Pfizer, "the chairmen or former chairmen of Citicorp, General Electric, [and] General Motors," and the Smith Richardson, Sarah Scaife, Weyerhauser, and John M. Olin Foundations (Himmelstein 1990, 147–148).

In turn, the Hoover Institution, the Heritage Foundation, and AEI achieved an important goal for corporate elites: they developed and widely promoted a neoliberal economic philosophy and set of concrete policy proposals that the Reagan administration largely adopted (Himmelstein 1990; Peschek 1987). For example, when President Reagan took office in 1981, the Heritage Foundation provided his administration with a 1,092-page report, titled *Mandate for Leadership*, that offered hundreds of specific policy proposals regarding virtually all areas of U.S. domestic and foreign policy. Many of these policy proposals were eventually adopted, including calls to reduce U.S. funding of the World Bank in favor of providing aid directly to developing nations and proposals to have developing nations "adopt free-market measures, reduce their state sector, and open their economies to the free flow of private foreign capital" (Peschek 1987, 105). Ultraconservative policy planning organizations also found demands made by developing nations in the 1970s for a redistribution of global wealth to be a serious threat to the "liberal world order" (Peschek 1987), a threat that Walden Bello et al. (1999) argue contributed to the Reagan administration's decision to adopt structural adjustment.

It is, of course, likely that the Reagan administration would have adopted at least some neoliberal policy proposals without the direct influence of ultraconservative policy planning organizations. However, in his first term, Reagan appointed 50 Hoover Institution affiliates, 36 Heritage Foundation affiliates, and 34 AEI affiliates to high level positions in his administration, with the result that his administration was, to a large degree, made up of representatives of these policy planning organizations (Himmelstein 1990; Peschek 1987). Moreover, his was the first Republican administration in which neoliberal ideologues such as these played a key role in shaping U.S. economic policy, making it difficult to conclude that the United States' abrupt neoliberal turn was not due, at least in part, to the incorporation of so many ultraconservative members of the policy planning network into his administration (Himmelstein 1990; Peschek 1987).

It is important to note, however, that Reagan's 1980 election victory came about in large part because he was able to form alliances not only with ultraconservatives but also with moderate conservative members of the power elite (Ferguson and Rogers 1981). As a result, his administration was staffed not only by neoliberal ideologues but also by less radical members of the power elite, such as Paul Volcker, Donald Regan, and Vice President George H. W. Bush, who all had strong ties to moderate conservative policy planning organizations such as the Council on Foreign Relations, the Committee for Economic Development, and the Trilateral Commission (of which Volcker was a member in 1979). It is therefore necessary to understand the position that moderate conservative policy planning organizations took at this time on neoliberal economic policy.

As Joseph Peschek (1987) demonstrates, and consistent with corporate elites' increasingly conservative political orientation (Himmelstein 1990), moderate conservative policy planning organizations supported many neoliberal economic policies in the late 1970s and early 1980s, including policies to radically liberalize trade and drastically reduce government intervention in the market in both developed and developing nations. For example, describing the Trilateral Commission's policy positions on global trade and government intervention in the late 1970s, Robert Cox states,

> The fundamental commitment of the [Trilateral] perspective is to an open world market with relatively free movement of capital, goods, and technology. Government interventions should be of a kind that support this goal, and such interventions as would impede it are to be condemned. *Powerful governments are to enforce this code of conduct upon weaker governments, using for this purpose especially the international organizations they control.* (Cox 1979, quoted in Peschek 1987, 84; emphasis added)

Thus, by the early 1980s moderate conservative and ultraconservative policy planning organizations supported similar economic policies, including the domination of developing nations by the world's wealthiest nations, but with a few important differences. Whereas moderate conservatives argued that their essentially neoliberal economic goals would be best achieved by (a) cooperating with other economically powerful countries and (b) using institutions such as the World Bank and IMF to enforce their will on developing nations, ultraconservatives tended to dismiss the importance of cooperation and to view international institutions such as the World Bank with great skepticism. In addition, while moderate conservatives generally believed that economic control over developing nations could be best achieved by ameliorating some of the worst effects that a globalized neoliberal order would have on these nations, ultraconservatives generally thought there was no need to ameliorate these effects (Peschek 1987).

These relatively minor differences in an otherwise similar vision of the global economic order led to some important contradictions in Reagan administration policy regarding developing nations and the global economy.[39] For instance, in accordance with some of the suggestions included in the Heritage Foundation's *Mandate for Leadership* report, the Reagan administration initially attempted to weaken the World Bank by cutting U.S. funding of the institution while simultaneously increasing direct U.S. aid to developing countries that enacted policies supported by the U.S.[40] However, when faced with the Latin American debt crisis, which could have produced a full-blown global economic crisis, the administration ended up strengthening the World Bank, not only to limit the geographic scope of the crisis and protect the world's financial architecture but also to promote neoliberal policies in the developing world. In doing this, the Reagan administration thereby used the World Bank, an institution that many neoliberals were skeptical of but that more moderate conservatives tended to support, to achieve a neoliberal goal that was supported by both groups but that was carried out in a manner that may have been harsher than moderate conservatives thought necessary or desirable.

Thus, the historical record demonstrates that members of the power elite, as funders of moderate conservative and ultraconservative policy planning organizations and important decision makers within the Reagan and Bush administrations, played key roles in (a) shaping the neoliberal economic policies adopted by the Reagan and Bush administrations and (b) convincing these administrations to use IMF and World Bank structural adjustment loans to address the Latin American debt crisis. Indeed, it is difficult to escape the conclusion that the Reagan administration was largely made up of members of the power elite and affiliates of a handful of moderate conservative and ultraconservative policy planning organizations that advocated these neoliberal policies from both inside and outside the administration. It is also difficult to escape the conclusion that when working for moderate conservative and ultraconservative policy planning organizations, these affiliates were essentially employees of the power elite. It is therefore reasonable to conclude that much of the blame for the social, economic, and environmental harm caused by World Bank and IMF structural adjustment loans resides with economic elites in the United States, who used undemocratic organizational mechanisms—the World Bank, the IMF, the U.S. policy planning network, and disproportionate access to White House decision makers—to achieve political and economic goals that benefited them to the detriment of hundreds of millions of people and the environment around the world.

The World Bank in the News

Given the important role the World Bank plays in harming people and the environment around the world and the important role the U.S. plays in shaping World Bank policy, one might expect these topics to receive a fair amount of attention in the U.S. news media. This, unfortunately, is not the case. In order to demonstrate this, I present a brief content analysis of all the news stories that appeared in the *New York Times* between January 1 and December 31, 2010, that included the term "World Bank." I chose the *Times* for this analysis because it is widely regarded as being the leading and most liberal mainstream newspaper in the United States (Mermin 1999) and therefore more likely than most other mainstream news sources to provide coverage critical of institutions such as the World Bank.

After deleting from my *New York Times* database all items that were editorials, obituaries, and theater and book reviews, and thus unlikely to be considered unbiased news reports by readers, I was left with 165 articles that mentioned the World Bank. Table 3.1 lists the topics covered by these stories and how many times each topic was covered in 2010. What is most immediately evident from looking at table 3.1 is how few of these 165 stories mention anything about how the World Bank is run (one story on World Bank governance), who most influences the World

Bank (zero stories), the kinds of conditions associated with World Bank structural adjustment loans (one story), or the negative social (two stories) and environmental (three stories) consequences of World Bank loans. Moreover, even these stories provide little information to the reader concerning these topics: the governance story discusses a reallocation of votes among World Bank member nations without discussing the Bank's board of directors, the vast power imbalances at the World Bank, or the fact that the redistribution of votes did nothing to change these power imbalances; the story that mentions the single loan condition only mentions it in passing; and the two stories that discuss the social consequences of some World Bank loans only begin to hint at these consequences.

The three environmental stories are somewhat more informative, with one story discussing a species of toad that nearly became extinct due to a World Bank dam in Tanzania and two of the stories discussing a World Bank loan for a coal-fired electricity plant in South Africa. Nevertheless, these are the only stories during the entire year (out of 165) that discuss the negative environmental consequences of World Bank activity; there is no attempt in any of these stories to connect the specific projects highlighted in the stories to other similar examples of World Bank activities that have harmed the environment; in the story on toads in Tanzania, the World Bank is essentially let off the hook for bearing any real responsibility for the problem; and in one of the stories on coal in South Africa, the World Bank makes a claim denying its environmental responsibility that is not countered in any way by quotes from other sources. Moreover, in the one story that does discuss the negative environmental consequences of a few World Bank projects in some detail, there is no direct way for the reader to know that the World Bank played an important role in promoting these projects.[41] Some readers may have correctly inferred this to be the case, but it is unlikely that many did. And even if they did, they could not be sure from the story that their inference was correct.

New York Times readers thus learned little from these stories about World Bank governance, structural adjustment, or the social and environmental consequences of World Bank lending policy. They did learn, however, that many important people have been affiliated with the World Bank (39 stories), that the World Bank collects vast amounts

TABLE 3.1. Content Analysis of *New York Times* News Stories on the World Bank in 2010

Topic	Number of occurrences[a]	Topic	Number of occurrences[a]
WB governance (how the WB is run)[b]	1	Which nations most influence the WB	0
Loan conditions	1	Social consequences associated with one or more WB projects	2
Environmental degradation associated with one or more WB projects	3	A person in the story is or was affiliated with the WB	39
WB data or evidence provided or referred to	59	Mention of WB investment or loan	12
WB provided debt relief	4	WB provided non-debt-relief assistance	9
A country or region is described as needing WB assistance or a WB assessment	3	Story describes recommendations made by the WB	10
Miscellaneous topics discussed[c]	38		

a The numbers in the columns add up to 181 because some of the stories covered multiple topics.
b WB refers to the World Bank.
c These 38 stories cover a diverse array of miscellaneous topics, such as gala events, that were too numerous to include as separate categories in the table. However, including them in the table as separate categories would not have changed the interpretation of the table provided in the main text.

of valuable data from around the world on a broad range of economic, social, political, and environmental topics (59 stories),[42] that the World Bank makes investments and provides loans in many countries around the world, usually for good causes such as fighting AIDS and poverty (12 stories), and that it provides debt relief and other forms of assistance to many nations (13 stories). Thus, the impression of the World Bank that *Times* readers gained in 2010 is of an organization that generates important knowledge, works with important people, and does good deeds around the world, an image that coincides very closely with that put forth by the World Bank itself (see, for example, Goldman 2005).

It should be clear, then, that readers of the *New York Times*, the leading and most liberal mainstream newspaper in the country, are left largely in the dark regarding the severe social, political, economic, and

environmental consequences of World Bank–sponsored activities and
the role that the U.S. and U.S. elites play in supporting these activities.
This ignorance, whether intentionally manufactured or not, diverts U.S.
citizens' attention away from the most consequential activities undertaken by the World Bank, from the role the U.S. and U.S. elites play in
shaping these activities, and from the global consequences of the kinds
of development projects and neoliberal policies the World Bank and U.S.
have pursued since the early 1980s. The *Times'* inadequate coverage of
the World Bank thus undermines democracy in the United States by
preventing U.S. citizens from knowing and understanding the policies
their elected and appointed officials pursue, the undemocratic manner in which these policies are developed, who these policies benefit,
and how they affect people around the world. Put differently, because
U.S. citizens are ignorant of these important matters, they are unable
to determine what is in their best interests or the best interests of their
country and, as a result, cannot vote or otherwise participate in policy
making effectively.

Conclusion

The theoretical model I set forth in chapter 2 (the IDE model) holds
that environmental degradation is to a significant degree the product
of organizational, institutional, and network-based inequality (OINB
inequality), which provides economic, political, military, geopolitical,
and ideological elites with the means to create and control organizational, institutional, and network-based mechanisms that they use to
achieve goals that are often socially and environmentally harmful. The
evidence presented in this chapter strongly supports the IDE model by
demonstrating that members of the U.S. power elite used their organizational and network-based power, which took concrete form in the
U.S. policy planning network and the network's and power elite's ties
to the White House, to get the U.S. to adopt structural adjustment as
a key policy tool for addressing the Latin American debt crisis and
spreading neoliberal orthodoxy throughout the developing world.
The evidence further supports the IDE model by demonstrating that
the U.S. used its organizational and financial power in the IMF and
World Bank to force these institutions to do its bidding during the

Latin American debt crisis as well as throughout much of these institutions' histories, with devastating human, social, and environmental consequences.

Clearly, the U.S. was not the only powerful nation to shape the events described in this chapter, and the IMF and World Bank are not without their own sources of power. There were also important differences in viewpoint across the moderate conservative and ultraconservative segments of the power elite. Nevertheless, in the United States moderate conservatives and ultraconservatives were essentially unified in their support of trade liberalization and reduced government intervention in the market, two key elements of structural adjustment; and the U.S. played a more important role than any other wealthy nation in shaping World Bank and IMF structural adjustment policies, with the IMF and World Bank essentially following the United States' lead regarding structural adjustment throughout the 1980s. Thus, it is safe to conclude that as part of their efforts to meet their neoliberal goals, economic and political elites in the U.S. used the World Bank, the IMF, and structural adjustment to shape the social, economic, and environmental policies of most developing nations, in the process severely harming people, communities, societies, and environments around the world. This, of course, is consistent with my theoretical argument.

Like structural adjustment, World Bank project-based loans have also negatively affected the lives and environments of hundreds of millions of people around the world, people who have had virtually no input into how these projects and loans have been planned and implemented.[43] To a significant degree project-based loans benefit both the World Bank, which has to make loans to ensure its survival and relevance, and World Bank lending officials, who are judged by their superiors on the basis of the number and size of the loans they put together (Goldman 2005; Meltzer 2000; Stiglitz 2002). However, these loans also provide tremendous benefits to (a) commercial banks and private investors (such as mining and dam construction companies), which often want the World Bank's seal of approval, organizational support, and/or insurance underwriting services before committing resources to specific development projects (Goldman 2005; UNCTAD 2000), and (b) large corporate contractors, such as Snowy Mountains Engineering Company, Lahmeyer International, and Louis Berger International, that end up receiving a large share of the money that the World Bank loans to developing countries (Goldman 2005; Paul et al. 2003).

It is beyond the scope of this chapter to investigate the degree to which these private banks, investors, and contractors influence World Bank decision making, but at a minimum it is clear that a small group of decision makers at the World Bank—loan officers, high-level officials, and wealthy nation executive directors—play a key role in providing or securing investment opportunities for and channeling public and private investment dollars to large, multinational corporations. These corporations, in turn, carry out development projects that harm local people and the environment, generate immense corporate profits, and make vast quantities of natural resources available to wealthy countries and corporations (see chapter 5 and Downey et al. 2010).

Since these socially and environmentally harmful projects usually fail to meet their economic development goals, rarely pull significant numbers of people out of poverty, and often promote the development of export-oriented economic sectors (such as agriculture and mining) that benefit wealthy nations and corporations but are poorly integrated into developing countries' national economies (see prior discussion and chapter 5; also see Bosshard et al. 2003; Rich 1994, 2009; Meltzer 2000; SAPRIN 2004; UNCTAD 2005), it is difficult to avoid the conclusion either that the World Bank is incompetent, which is difficult to believe, or that the goal of these projects is to benefit corporations and the wealthy nations that run the World Bank. Like structural adjustment, then, World Bank project-based loans are properly viewed as an elite-controlled organizational mechanism that is used to promote elite interests at the expense of developing nation citizens and the environment around the world. This, of course, is also consistent with my overarching theoretical argument.

In addition to supporting my overarching argument, the evidence presented in this chapter also supports my claim that elite-controlled organizations, institutions, and networks are a fundamental source of social and environmental degradation because they provide elites with the means to (a) monopolize decision making power, (b) shift environmental and non-environmental costs onto others, (c) inhibit the development and/or dissemination of environmental knowledge, attitudes, values, and beliefs, (d) restrict the ability of non-elites to behave in environmentally sustainable ways, (e) divert the public's attention away from what elites are doing so that their actions will not be scrutinized, questioned, or challenged, and (f) frame what is and is not considered to be good for the environment. For instance, despite developing and enforcing policies that affect hundreds of millions, if not billions, of people around the world, the World Bank and IMF have absolutely no democratic accountability. Only a handful of wealthy countries have any real power in these institutions, with the U.S. having by far the most power, and in these wealthy countries, it is the finance ministers, treasury secretaries, and central bank governors, who are all closely tied to the finance industry rather than the citizenry, who most influence the World Bank and IMF (Stiglitz 2002). As an example of this, it was Federal Reserve Chairman Paul Volcker and U.S. Treasury Secretaries Donald Regan, James Baker, and Nicholas Brady who played the key roles in determining how the World Bank and IMF responded to the Latin American debt crisis.

The historical record therefore demonstrates that in the 1980s World Bank and IMF policy was set not by a broad coalition of actors from around the world but by representatives of the U.S. power elite and finance industry, who until at least 1989 worked for an administration whose high level staff were drawn largely from just three ultraconservative think tanks. Economic elites in the U.S., with the aid of U.S. political elites, were thus able to use the World Bank, the IMF, the U.S. policy planning network, and power elite ties to the White House to *monopolize U.S. and global decision making power* to such a degree that this small group of people and organizations was able to make decisions that affected the lives of hundreds of millions, perhaps billions, of people around the world. In making these decisions, economic and political elites in the U.S. (and other wealthy nations) undermined the sovereignty of nearly every developing nation in the world, as well as the rights of citizens in democratic developing nations to influence their nations' policies, with disastrous consequences for these nations' economies, communities, and environments. And because the U.S. and other wealthy nations still exert tremendous influence over the IMF and World Bank, which, in turn, continue to impose structural adjustment and large development projects on the world's developing nations, this basic principle still holds true today.

It is important to note, however, that the ability of economic and political elites in the U.S. to monopolize decision making power in this way did not affect only developing nation citizens. Because democracy can only operate when every citizen and social group in a country has a relatively equal opportunity to influence her or his nation's political decisions and public policy, the disproportionate influence that members of the power elite and a handful of policy planning organizations had on U.S. policy in the 1980s also undermined democracy in the United States, where only a small group of elite interests were represented in the policy making process regarding the Latin American debt crisis.

It is thus clear that the elite-controlled organizational mechanisms described in this chapter spread neoliberal policies to the developing world in large part by restricting decision making power to a privileged few elite representatives. These elite-controlled mechanisms also shifted many of the most severe social and environmental costs of global economic growth and wealth generation to citizens of the developing world. These costs are not borne by members of the power elite or by political elites in the U.S. and other wealthy nations; and for the most part they are not borne by wealthy nation citizens either. Thus, in using the IMF and World Bank to achieve their economic and political goals, economic and political elites in the world's wealthiest nations have ensured (intentionally or not) that poor, working class, and middle class people in developing nations bear the greatest social, economic, and environmental costs of global economic development and capital accumulation. They have also ensured that the most onerous effects of economic growth, capital accumulation, and wealthy nation economic policy are geographically removed, and thus hidden, from wealthy nation citizens, thereby diverting these citizens' attention away from these effects and helping to strengthen these citizens' support for economic growth, trade liberalization, free markets, and neoliberalism.

As we have seen, World Bank knowledge generation and suppression activities also help to divert wealthy nation citizens' attention away from the harmful effects of World Bank policy by inhibiting the development of accurate knowledge regarding the social and environmental consequences of this policy and by framing what is and is not considered to be pro-environmental behavior, policy, and development. The World

Bank's ability to portray its project and loan interventions in a positive social and environmental light is, as I previously discussed, also reinforced by the inability or unwillingness of the major U.S. news media, or at least the *New York Times*, to cover these interventions in anything but the most cursory, superficial, and positive manner possible.

However, without accurate knowledge of the consequences of World Bank policy or of the role the U.S. plays in shaping this policy, U.S. citizens cannot behave in socially and environmentally responsible ways even if they want to do so. They are, for example, unlikely to exert pressure on their elected representatives to make World Bank activities more socially and environmentally responsible or to support those who do try to exert such pressure if they know nothing about the World Bank or if they believe the World Bank is an agent for positive social and environmental change around the world. Without accurate knowledge of World Bank activities, they are also unlikely to alter their consumption behavior so as to avoid purchasing goods derived from World Bank–supported resource extraction projects, thereby undermining the utility of the consumption-oriented environmental and social change strategies I discussed in chapter 1.

It is thus apparent that the evidence presented in this chapter supports all six of the hypotheses I set forth in chapter 2. In particular, the evidence demonstrates that economic and political elites in the U.S. used the organizations, institutions, and networks highlighted in this chapter to monopolize decision making power, shift environmental and non-environmental costs onto others, limit non-elites' pro-environmental behavior, inhibit the generation and dissemination of environmental knowledge, divert the public's attention away from what elites are doing, and frame what is and is not considered to be pro-environmental policy and development.

The clear intent with which economic and political elites in the U.S. used these organizations, institutions, and networks to force developing nations to adopt wide ranging neoliberal policies also demonstrates that globalization and neoliberalism are not natural and inevitable forces but are rather the political achievements of a very small group of economic and political elites in the world's wealthiest nations, who have worked very actively and deliberately over the past four decades to achieve their neoliberal goals.

Finally, the fact that these elites used the undemocratic organizations, institutions, and networks highlighted in this chapter to bail out the world's largest banks and provide wealthy nations and corporations with access to natural resources, markets, and investment opportunities demonstrates that elite-controlled organizations, institutions, and networks are an essential component of the large-scale social structures, or structures of accumulation, that make capital accumulation possible.

The evidence presented in this chapter thus confirms my overall theoretical argument, the six hypotheses I derived from this argument, and the argument I made at the beginning of this chapter regarding the purported inevitability of globalization.

I turn now to an examination of inequality and power in the global agriculture industry.

4

Modern Agriculture and the Environment

When most people in the U.S. think about farming, a rather idyllic picture of green pastures, red barns, and farmers in overalls comes to mind. Absent from this picture are images of farmers struggling, and often failing, to make ends meet (45% of all U.S. farms lost money in 2010); of farmers using vast quantities of pesticides and fertilizers that taint our food, end up in our nation's waterways, and poison farmers, farmworkers and farming communities; of thousands of animals packed into inhumane production facilities where they are fed excessive quantities of antibiotics; or of corporations that manipulate agricultural commodity markets, dictate farming practices, and routinely sue farmers for contract violations. Nor are most people aware of the key role that farming plays in producing environmental crises such as ocean acidification, severe soil erosion, freshwater scarcity, and global warming.

Farming has, of course, always been a difficult profession, and it has often been associated with serious environmental problems. However, since World War II the social and environmental problems associated with farming have increased dramatically as agriculture has been transformed from a relatively decentralized and highly labor-intensive activity to a highly integrated, corporate dominated, and capital-, resource-, and technology-intensive industry dependent on heavy machinery, petrochemicals, and fossil fuels (Heffernan 2000).

This transformed industry and the production technologies that it uses not only play a key role in shaping many local, regional, and global environmental crises, they also threaten to undermine the basis of increased agricultural yields (Carolan 2011; Hawken et al., 1999) while simultaneously producing poverty and hardship for peasants and farmers around the world and great wealth and power for the world's largest food retailers and agribusiness firms (Hendrickson et al. 2008). Moreover, despite dramatic increases in food production since World War II and global efforts to eradicate hunger, 925 million people, or 16% of the developing world, still experienced chronic hunger in 2010, with hunger in developing nations contributing to widespread stunted growth among children and to just under three million childhood deaths that year alone (FAO 2010b).

Given the extremely serious social and environmental problems associated with modern agriculture (described in detail later in the chapter) and the fact that modern agricultural techniques continue to spread rapidly around the world, it is imperative that we understand not only the negative consequences of our modern system of food production but also the social forces and structures that have created and maintained this system. To that end, this chapter (a) examines eight mechanisms that economic and political elites have used to achieve U.S. agriculture industry goals, including oligopoly and oligopsony power in agricultural commodity chains, liberalized agricultural trade, international property rights regimes, and decision making rules at the World Trade Organization (WTO) that favor the U.S. and Europe, and (b) asks whether these undemocratic and elite-controlled mechanisms have played an important role in producing human misery, social and environmental harm, and corporate wealth accumulation.

I begin with a brief discussion of some of the most serious social and environmental problems associated with modern industrial agriculture. I then link these problems to the structure of the modern agriculture industry and in a series of case studies explain how the eight undemocratic and elite-controlled mechanisms highlighted in the chapter have simultaneously structured this global industry, harmed individuals, societies, and the environment, and provided elites with the means to achieve their capital accumulation goals.

The first of these case studies examines how large agribusiness firms use oligopoly and oligopsony power in agricultural commodity chains to appropriate value from farmers, shape and restrict the agricultural techniques that farmers use, and limit the food choices available to consumers. The second case study examines elite-controlled policy planning networks in the U.S. and the role they have played in shaping U.S. agricultural trade policy, the WTO's Agreement on Agriculture, and U.S. support for the trade-liberalizing Agreement on Agriculture. Finally, the third case study examines how the mobilization of elites through a set of social and organizational networks that linked corporate leaders in the U.S., Europe, and Japan to each other and to their respective governments resulted in the adoption of the WTO's socially and environmentally harmful Agreement on Trade-Related Aspects of Intellectual Property Rights (TRIPs).

As we shall see, these case studies provide strong support for the argument that environmental degradation and social harm are to a significant degree the product of undemocratic organizational, institutional, and network-based mechanisms that elites use to achieve their capital accumulation goals and, thus, the result, more generally, of OINB inequality. As we shall also see, these case studies support three of the six hypotheses I set forth in chapter 2, in particular, my predictions that elite-controlled organizations, institutions, and networks are a fundamental source of social and environmental degradation because they provide elites with the means to monopolize decision making power, shift environmental and non-environmental costs onto others, and restrict the ability of non-elites to behave in environmentally sustainable ways. In providing empirical support for these three hypotheses, these case studies also demonstrate that the elite-controlled mechanisms examined in this chapter play a theoretically similar role in producing social and environmental harm as do the elite-controlled mechanisms described in the preceding and following chapters, thereby providing additional support for my theoretical model.

I turn now to a discussion of modern agriculture's social and environmental consequences.

The Social and Environmental Consequences of Industrial Agriculture

Modern industrial agriculture is a system of intensive, large-scale food production characterized by a specific set of production technologies and organizational and institutional attributes, including motorized farm equipment; monocropping (growing the same crop in the same fields year after year); large, often absentee-owned, farms; immense confined animal feeding operations (CAFOs); heavy reliance on irrigation, antibiotics, growth hormones, chemical inputs, and fossil fuels; extremely high levels of corporate concentration; highly liberalized markets; and the integration of farms and ranches into vertically structured commodity chains dominated by a small number of very large retailers and agribusiness firms (Gurian-Sherman 2008; Heffernan 2000; Hendrickson et al. 2008).

As many observers have noted, the widespread diffusion of these technologies and organizational/institutional attributes after World War II helped to dramatically increase food production in the second half of the 20th century.[1] The diffusion of these technologies and attributes also undermined the ecological viability of agriculture in many parts of the world and produced many severe social, environmental, and public health problems both locally and globally. For instance, although global food production more than doubled between 1965 and 2005 (Khan and Hanjra 2009), the amount of pesticides (herbicides, fungicides, insecticides, and bactericides) and synthetic fertilizers applied to crops and croplands also increased dramatically during this time period, due in large part to the nutritional requirements of high-yield crops (Shiva 1991), the negative effect that intensive farming has on soil nutrient quality (Hawken et al. 1999), and the extreme vulnerability of large monoculture fields to pests, weeds, and diseases (Gurr et al. 2004). As a result, by the end of the first decade of the new millennium, global agriculture used approximately three million metric tonnes of pesticides (Pimentel 2009) and 175 million metric tonnes of fertilizer (FAO 2008a) per year.

Of course, these pesticides and fertilizers do not stay on the farm. Pesticides that are sprayed onto fields are often blown onto nearby communities, causing chronic and acute pesticide poisoning (J. Harrison 2011); and nitrogen and phosphorous from fertilizer enter waterways in significant volumes, where they elevate the water's nutrient content, in extreme cases producing algal blooms that use up the water's oxygen. This can create dead zones, such as a 6,000–7,000 square mile area in the Gulf of Mexico that forms every spring and summer, where few organisms are able to live, harming not just the environment but also local economies dependent on recreational and commercial fishing (Bruckner 2011).

The massive use of pesticides in agriculture is also quite problematic. Each year, for instance, pesticide use results in approximately 26 million cases of poisoning and 220,000 deaths worldwide (Carolan 2011).[2] Moreover, chronic exposure to pesticides can cause cancer, polyneuropathy (a neurological disorder that causes nerves throughout the body to malfunction), dermatitis, behavioral changes, and respiratory and reproductive problems (Singh and Gupta 2009; Pimentel 2005).

Widespread pesticide use also results in pesticide resistance among targeted species and can lead to groundwater contamination, the poisoning and killing of birds, fish, and other animals, and the loss of pollinator species and species that protect crops against harmful plants and insects (Carolan 2011; Pimentel 2005). Moreover, pesticide use is at best only partially successful: despite spending about $40 billion per year on pesticides globally, between 35% and 42% of the world's crops are still destroyed each year by insects, weeds, and plant pathogens (Pimentel 2009).

Modern agricultural techniques and the clearing of land for small-scale farming also severely erode the soil, with at least 75 billion tons of soil lost each year due to agriculture (Blanco and Lal 2010; Pimentel 2006), nearly "80% of the world's agricultural land suffer[ing] moderate to severe erosion, . . . [and] an estimated 10 million [hectares] of cropland worldwide [being] abandoned [each year] due to . . . erosion" (Pimentel 2006, 123).[3]

Soil erosion and the use of heavy machinery, pesticides, and fertilizers also deplete nutrients from the soil (Pimentel 2006) and reduce the "ability of soil bacteria, fungi, and other tiny organisms to cycle nutrients, fight disease, and create the proper soil texture and composition to protect roots and hold water" (Hawken et al. 1999, 193). These factors, in turn, significantly decrease the overall productivity of farmland, such that by the early 1990s corn productivity had declined by 9%–18% in Indiana, by 25%–65% in Georgia's southern Piedmont region, and by up to 80% in the Philippines, with annual crop productivity losses due to erosion in the U.S. in the late 1990s estimated to be $37.6 billion (Pimentel 2006, 127).

In addition to eroding and degrading the soil, modern agriculture also uses vast quantities of freshwater, particularly for irrigation, which consumes roughly 80% of all the water used globally for all purposes (Hanjra and Qureshi 2010). This not only jeopardizes water security for many people, it also threatens agricultural production in many regions of the world by rapidly decreasing freshwater supplies, increasing the salinity of the soil, and producing waterlogged soil (Khan and Hajra 2009). Indeed, by the late 1990s roughly one-tenth of the world's irrigated farmland had experienced serious productivity declines due to soil salinization and waterlogging (Korten 2001), with an additional two million hectares per year experiencing severe or total productivity losses due to these factors (Khan and Hanjra 2009). These declines, in turn, have "offset[] a significant proportion of the gains in agricultural productivity achieved through the Green Revolution. . . . One estimate, for example, shows that the degradation of irrigated lands used to produce rice and wheat in the Punjab region [of] Pakistan reduced the gains made by breeding and infrastructural and educational investments by approximately 33 per cent" (Murgai et al. 2001, as summarized in Carolan 2011, 121).

Moreover, pesticide and fertilizer use, soil erosion, and irrigation also negatively affect agricultural productivity around the world, with the annual growth rate of global wheat and rice production declining from 5% to 2% and 3% to 1%, respectively, between 1980 and 2005 (FAO 2011) and overall growth in global agricultural yields estimated to be declining by roughly 1.5% per year (Hanjra and Qureshi 2010). It thus appears that modern agriculture's attack on the environment will play a key role, along with population growth, lack of freshwater, and unequal access to land and food (FAO 2011; GRAIN 2008; Hanjra and Qureshi 2010), in producing severe food shortages over the next 50–100 years, thereby threatening food security for billions of people around the world (FAO 2011).

Further complicating matters, the agriculture industry is one of the world's major contributors to the climate change crisis due to its reliance on fertilizers, petroleum-based pesticides, fossil-fuel powered machinery, and farming practices that release vast quantities of soil-based carbon into the atmosphere. The widespread clearing of forests for crops and grazing, the use of petroleum-based global distribution networks, and the vast quantities of carbon dioxide, methane, and nitrous oxide produced by the billions of livestock inhabiting the planet also contribute significantly to climate change (FAO 2008b; Goodland and Anhang 2009; Hawken et al 1999; LaSalle and Hepperly 2008; Lokupitiya and Paustian 2006; Steinfeld et al. 2006). Livestock production, for example, is responsible for roughly 9% of all anthropogenic carbon dioxide emission, 37% of all anthropogenic methane emissions (which have 23 times the global warming potential of carbon dioxide), and 65% of all anthropogenic nitrous oxide emissions (which have 296 times the global warming potential of carbon dioxide) (Steinfeld et al. 2006, xxi). Moreover, the agriculture industry *as a whole* produces about one-third of all global carbon dioxide emissions and more than half of all global methane emissions (LaSalle and Hepperly 2008; FAO 2010a).

These emissions pose a key threat to future agricultural productivity. The International Food Policy Research Institute (IFPRI) notes, for instance, that the higher temperatures and changing precipitation patterns associated with climate change will

> reduce yields of desirable crops, . . . increase the likelihood of short-run crop failures and long-run production declines, . . . [and] result in . . . price increases for the most important agricultural crops—rice, wheat, maize, and soybeans. Higher feed prices will result in higher meat prices. As a result, climate change will reduce the growth in meat consumption slightly and cause a more substantial fall in cereals consumption. . . . Calorie availability in 2050 will [thus] not only be lower than in [a] no-climate-change scenario—it will actually decline relative to 2000 levels throughout the developing world, . . . [thereby] increase[ing] child malnutrition by 20 percent relative to a world with no climate change. (Nelson et al. 2009, vii)

Modern agriculture also relies on a very narrow range of crops. It has been estimated, for example, that "about 75 per cent of plant genetic diversity has been lost as farmers worldwide have abandoned their local varieties for genetically uniform varieties that produce higher yields [but] under [very specific] conditions" (De Schutter 2009, 14).

This dramatic decline in crop genetic diversity is exceedingly dangerous. In particular, by narrowing the pool of plant traits from which farmers can draw, reduced genetic diversity makes the world's major crops increasingly vulnerable to pests, disease, and climate change, with potentially disastrous consequences not only for farmers but also for the world's food supply.

Somewhat perversely, then, the technologies used to increase agricultural productivity in the 20th century now threaten to undermine worldwide food production by degrading, and sometimes destroying, the environmental and genetic conditions necessary for growing crops and raising livestock (Foster 1999). These technologies, and the agricultural system that produced them, also harm farmers and farming communities around the world. One of the most important ways in which they do this is by keeping the costs of farm inputs (seeds, pesticides, fuel, etc.) high and the prices farmers receive for their goods low, resulting in decreasing farmer incomes, increasing levels of farmer debt, and a shift toward ever larger farms that make up for decreased profit margins by producing extremely large quantities of agricultural goods (Carolan 2011). In 2009, for example, the largest 2% of farms in the U.S. were responsible for 39% of all U.S. farm production (White and Hoppe 2012), and in 2010 45% of all farms in the U.S. lost money (Park et al. 2011).

Contributing to their inability to make a profit, small and medium-size farms around the world are finding it increasingly difficult to access the markets they need to sell their goods (Dolan and Humphrey 2001, 2004; Hauter 2009). The choices farmers make about the kinds of goods they produce and the techniques they use to produce these goods are also being increasingly shaped and dictated not by themselves or the market but by large food retailers and agribusiness firms whose interests differ greatly from those of farmers and consumers (Dolan and Humphrey 2001, 2004; International Trade Centre 2011; Downey and Strife 2010; Heffernan 2000; Hendrickson et al. 2008).

Of course, declining incomes, increased debt, and reduced control over their lives do not only produce economic hardship. In many instances, these factors also produce severe depression and desperation, as evidenced by alarmingly high rates of farmer suicide around the world (Behere and Bhise 2009; Gorelick 2000; Sainath 2009). In India, for example, official statistics show that more than 290,000 farmers killed themselves between 1995 and 2014, though the actual number is likely to be much higher due to unreliable population counts and inadequate reporting (Barry 2014; Joseph et al. 2003; Sainath 2009); and in India, Sri Lank, Canada, England, Australia, and the U.S, suicide rates are higher among farmers than among the general population (Behere and Bhise 2009).

The question, of course, is how and why did we create and how and why do we perpetuate a system of global food production that so severely degrades the environment, so severely undermines its own ecological viability, and so severely impoverishes farmers; a system that produces enough calories for every person in the world while leaving nearly one billion people hungry and malnourished? It is, of course, impossible to fully answer these questions in a single chapter. Thus, in the following pages I provide a detailed answer to the question of how and why we perpetuate this system, focusing initially on the role of oligopoly and oligopsony power in agricultural commodity chains.

Oligopoly and Oligopsony Power in Agricultural Commodity Chains

Commodity chains and oligopoly and oligopsony power are two important mechanisms used by agriculture industry elites to extract value from farmers and suppliers and ensure the widespread adoption and continued use of farming techniques that benefit these elites. Commodity chains are organizational networks, composed primarily of competing and allied companies but also of individuals and other organizations, through which raw materials and semifinished goods travel as they are transformed into finished products that are eventually sold in the marketplace. For any specific product, commodity chains connect all the stages of and all the organizations and individuals involved in the production and resource extraction process to each other and to the final consumer.

In some commodity chains, multiple stages of production are housed in each firm, while in other commodity chains, most of the different stages are carried out by different firms. When the different stages of extraction and production are carried out by different firms or actors, these firms and actors must enter into market exchanges with each other so that the raw materials or semifinished products they handle can move up the chain. In many, if not most, cases these exchanges are carried out by firms and actors with vastly different levels of market power.

The terms *oligopoly* and *oligopsony* refer to two specific types of situation in which market power between commodity chain actors is highly unequal. An oligopoly exists when, in any given market, there are numerous buyers but only a small number of sellers (referred to as oligopoly firms), while an oligopsony exists when, in any given market, there are numerous sellers but only a small number of buyers (oligopsony firms). A monopoly or monopsony exists when there is only a single seller or buyer.

Oligopoly and oligopsony firms tend to exert great power in commodity chains *and* markets, generally distorting these chains and markets to their benefit by eliminating or reducing free and fair competition between themselves and other actors. They do this in several ways, four of which I discuss here. First, because these firms have many buyers or sellers to choose from, while potential buyers and sellers have only a handful of firms to do business with, these buyers and sellers are often forced to accept the terms of exchange dictated by the oligopoly or oligopsony firm, especially in situations where the market is structured such that the buyer or seller has access to only a single oligopoly or oligopsony firm.[4] Second, because oligopoly and oligopsony firms are generally quite large, and almost always much larger than their buyers and sellers, they usually have much more market information than do their buyers and sellers, giving them an unfair advantage in setting the terms of market exchange. Third, because of their great size, oligopoly and oligopsony firms are often able to unfairly undercut market competition and shape or set market prices by flooding markets with their goods or selling their goods at below cost, making up for their losses with money from their cash reserves or from their sales in other markets (this power can be used to eliminate firms in the same stage of production as the oligopoly or oligopsony firm or to shape the conditions in which the firm's buyers and sellers operate).[5] Fourth, because of their size and financial resources, oligopoly and oligopsony firms are often able to influence government policies in ways that produce even more favorable market outcomes for them.

Because oligopoly and oligopsony power can exist only when a small handful of firms control a disproportionate share of the market, the existence of potential oligopolies and oligopsonies is usually determined by measuring the share of the national or global market controlled by the four or five largest firms in the market, with markets generally held to be highly concentrated when four firms collectively control at least 20% of the market and to be extremely concentrated, and exhibiting oligopolistic or oligopsonistic conditions, when four firms control more than 40% of the market (Heffernan 2000; Wise 2005). It is important to note, however, that because this measure of (horizontal) market concentration is generally determined based on the largest companies' shares of national sales, it likely misrepresents the level of *oligopsony* power that exists in at least some markets. Wenonah Hauter (2009) argues, for example, that oligopsony power can be exerted at much lower levels of horizontal concentration than oligopoly power because firms with relatively low levels of national sales may still be the only buyer in a particular region of the country. Moreover, farmers are particularly vulnerable to oligopsony power because the goods they sell are generally perishable, often need to be sold during a fairly short time window, and in many cases are expensive to transport long distances (Hauter 2009).

As is true in many sectors of the U.S. economy, the agriculture industry in the United States is very highly concentrated, with the four largest firms in the beef, pork, broiler (chicken), and turkey slaughter sectors controlling 82%, 63%, 53%, and 58% of their respective U.S. market in 2009 and the four largest animal feed, soybean processing, and rice milling firms controlling 44%, 85%, and 55% of their respective U.S. market in 2007 (James et al. 2012). In 2007 the three largest flour milling and wet corn milling firms controlled 52% and 87% of their respective U.S. market (James et al. 2012), and in that same year the four largest nitrogen fertilizer, phosphate fertilizer, farm machinery, and agricultural implements firms controlled 49%, 83%, 59%, and 55% of their respective U.S. market (U.S. Census Bureau 2007). Finally, in 2008 the four largest supermarket chains in the U.S. controlled 49% of the U.S. grocery market, though in some major cities the share of the market controlled by the largest supermarkets was much higher than this (Carolan 2011).

The U.S. crop seed industry is similarly concentrated, with Monsanto and Pioneer controlling 60% of the U.S. corn and soybean seed market in 2003 (Hendrickson et al. 2008), Monsanto and DuPont controlling 58% of the U.S. corn seed market in 2007 (Hendrickson and Heffernan 2007), and Monsanto "holding patents on genetic material found in 80% of the corn and 93% of the soybeans grown in the U.S. in 2009" (Hauter 2009, 14).

The situation is not much different at the global level, where in 2007 the world's four largest proprietary seed companies controlled 53% of the global market (82% of global commercial seed sales are proprietary),[6] the world's five largest agrochemical firms controlled 68% of the global market, the world's seven largest fertilizer companies controlled almost the entire global market, the world's four largest grain traders handled between 75% and 90% of the global grain trade, and the top 10 supermarket chains controlled 40% of the sales of the world's 100 largest grocery chains, which in turn accounted for 35% of all global retail grocery sales (Walmart alone was responsible for 10% of the top 100's sales) (ETC Group 2008; Renwick et al. 2012).

It is thus apparent that large agribusiness firms exert significant market power in agricultural commodity chains. The power of these firms is enhanced, moreover, by the fact that many of them operate in multiple stages of the commodity chain (they are vertically integrated). Table 4.1 demonstrates this for the U.S. by listing firms that were among the four largest companies in 2007 or 2009 in two or more stages of the chain. The table shows, for instance, that Cargill was among the top four firms in beef production (feedlots), beef, turkey, and hog slaughtering, animal feed production, flour and wet corn milling, and soybean processing.[7] It was also one of the world's four largest global grain traders. Similarly, Archer Daniels Midland (ADM) was one of the United States' four largest soybean processors, animal feed producers, and flour, wet corn, and rice millers, as well as one of the world's four largest grain traders. And in 2007 (see table 4.2) four of the world's seven largest proprietary seed companies were also among the world's seven largest agrochemical companies. Monsanto and DuPont, for example, were the world's largest and second largest proprietary seed companies and the world's fifth and sixth largest agrochemical companies, while Syngenta was the world's second largest agrochemical company and its third largest proprietary seed company. Moreover, the world's fifth largest proprietary seed company, Land O'Lakes, was one of the four largest animal feed producers and dairy processors in the U.S. at that time.

Further eroding competition in agricultural markets, the world's largest agribusiness firms sometimes enter into formal alliances with each other through joint ventures, partnerships, contracts, and formal agreements. In 1999, for instance, Cargill and Monsanto formed a joint venture to develop, market, and distribute genetically engineered products for the grain processing and animal feed industries (Monsanto 1998; *St. Louis Business Journal* 2006), giving these companies oligopoly and oligopsony power throughout the entire agricultural commodity chain (Heffernan et al. 1999). And in recent years giant seed and agrochemical firms have also formed important alliances with each other, including alliances created in 2007 and 2008 between Monsanto and BASF,

TABLE 4.1. The Largest Agribusiness Firms in the U.S., 2007 and 2009

	Cargill	Tyson	JBS	Smithfield	Land O'Lakes	ADM	Bunge
Beef slaughter	X	X	X				
Beef production	X	Link to a top 4	X				
Hog slaughter	X	X	X	X			
Hog production				X			
Broiler slaughter	X	X	X				
Turkey slaughter	X			Part owner			
Dairy processor					X		
Animal feed	X				X	X	
Flour milling	Part owner					X	
Wet corn milling	X					X	
Soybean processor	X					X	X
Rice milling						X	
Global grain trade	X					X	X

Source: James et al. 2012, table 1.

TABLE 4.2. The World's Largest Seed and Agrochemical Firms, 2007

	Global rank	
	Seed	Agrochemical
Monsanto	1	5
DuPont	2	6
Syngenta	3	2
Land O'Lakes[a]	5	—
Bayer	7	1

Source: ETC Group 2008; James et al. 2012, table 1.
[a] Land O'Lakes was one of the four largest U.S. dairy processors and animal feed producers in 2007.

Monsanto and Dow, Monsanto and Syngenta, and Syngenta and DuPont (ETC Group 2008).

The result of all this is that large agribusiness firms generally exert oligopoly and oligopsony power at multiple stages of the agricultural commodity chain, with large supermarket chains also exerting tremendous oligopsony power over their suppliers. As I describe in the following sections of the chapter, this power gives agribusiness firms and food retailers the ability to strongly shape, and in many cases dictate, not only the prices that farmers receive for their goods but also the agricultural techniques that farmers use to grow crops and raise livestock. One of the chief ways this occurs is through the manipulation of markets.

Market Manipulation

Marketing and production contracts are two important tools that oligopsony firms use to manipulate agricultural markets and accumulate capital. Marketing contracts are contracts in which famers agree to supply an agribusiness firm with an agricultural commodity at a future date. Production contracts require farmers to provide agricultural or farming services for an agribusiness firm:

> Buyers use marketing contracts to secure a reliable supply of the input they process, generally livestock. The farmers make production decisions with limited oversight by the contract buyer and own the commodity they are producing. Farmers are paid based on a formula price (either

a fixed baseline price or a price tied to the spot or futures market) that can vary based on volume. Buyers develop these input streams (known as captive supplies) because they allow the buyer to exercise power over the seller without competing with other potential buyers on open or spot markets at the time when they need the livestock. Production contracts [on the other hand] pay farmers for the service of raising the crop or livestock, not the crop or livestock itself. The agricultural processing company [owns and] delivers the inputs (seed, feed, young livestock, transportation, etc.) to the farmer and then picks up the farm goods when the production is complete. (Hauter 2009, 9)

In 2008 marketing and production contracts covered, as a percentage of the dollar value of production, 39% of all U.S. agricultural production, 90% of poultry and sugar beet production, 68% of hog production, 54% of dairy production, 29% of cattle production, 45% of rice production, 38% of fruit production, 39% of vegetable production, and 26%, 25%, and 23% of corn, soybean, and wheat production, respectively: contracts for corn, soybean, and wheat tend to be marketing contracts, while contracts for poultry and hogs are generally production contracts (MacDonald and Korb 2011).

Contracts distort farm gate prices (the prices farmers receive for their goods) in a number of ways. First, because contract sales occur outside the regular cash, or spot, market, prices are not determined solely in the open market, and those prices that are determined in the open market likely "reflect a nonrepresentative set of transactions, making the reported prices an inaccurate reflection of activity" (MacDonald et al. 2004, quoted in Hauter 2009, 10). Second, because contract prices are not recorded publicly, farmers lack information about the prices other farmers are receiving for their crops and livestock, making it more difficult for them to determine whether they are receiving a fair market or contract price for their goods. Third, because contract sales pull commodities off the cash market, it becomes easier for commodity traders and livestock processors to influence the prices farmers receive in the cash market. For instance, because meatpackers directly own large numbers of "off-market" cattle, they can slaughter them when the cattle cash price is high, thereby reducing aggregate market demand, and purchase on the cash market when prices are low (Hauter 2009).

In addition to distorting prices through the use of contracts, agribusiness firms also influence prices by manipulating commodity futures markets (Carstensen 2004). Hauter explains:

> Buyers of farm products that have prices based on the prices on the commodity futures market have an incentive to manipulate the futures price to impact the actual purchase prices. For example, the cash or spot price for live cattle is influenced by the price for live cattle commodity futures contracts, so meatpackers can participate in the futures market to influence the cash price they pay for cattle. . . . This may be especially true of thinly traded commodities, like hogs, where the futures market represents a tiny share of the national hog market, but hog contract prices are based on a the commodities futures prices. Beef packers [also] buy and sell . . . futures contracts and some have seats on the futures exchanges, which allows these companies to exert market pressure on the futures contract price of cattle, which in turn impacts the spot price, and vice versa. (2009, 8)

Moreover, spot markets are often very *un*competitive. Feedlot cattle auctions, for example, often attract only one or two of the major beef packers, and 57% of all feedlots sell to only a single buyer, making it difficult or impossible for ranchers, who cannot afford to haul cattle to distant locations, to get a fair market price for their cattle (Hauter 2009).

In all these cases farmers are forced to accept lower prices than they would accept in a fully competitive market because they have little choice but to do business with one or a few large buyers who have more information than they do about the prices other farmers are getting for their goods, who can source their inputs from multiple farmers using either contracts or highly distorted spot (cash) markets, and who can manipulate prices in interconnected contract, futures, and spot markets.

Low farm gate prices translate into higher profits for agribusiness firms and lower profits and profit margins for farmers, which in turn encourage farmers to increase both production and farm size. Consistent with this, between World War II and 2006 the amount of money farmers received for every dollar consumers spent on food declined from roughly 50 cents to 19 cents (Carolan 2011; Domina and Taylor 2010), while the years 1991–2009 saw farms with more than $1 million in sales increase their share of U.S. farm production from 21% to 39% (White and Hoppe 2012).

Agribusiness firms also prefer doing business with large farms because working with a small number of large farms reduces these firms' transaction costs (Hauter 2009). As a result, in 2008 only 6.6% of farms with less than $250,000 in sales had production or marketing contracts, while 53.2% of those with sales between $250,000 and $499,999 and 69.9% of those with sales over $1 million had such contracts (MacDonald and Korb 2011). It is thus clear that market manipulation and oligopsony power do not simply lower farm gate prices. They also shape the structure of the U.S. farm system by favoring large farms over small and by influencing farmers' decisions about optimal farm size and where to market their crops and livestock.

Another way in which farm gate prices and farmer behavior are influenced by oligopsony firms is through the use of government subsidies that agribusiness firms and retailers, due to their economic and political power, have successfully lobbied for since the 1970s (Billig and Wallinga 2012; Pollan 2003; Tillotson 2004). Although these subsidies are provided directly to farm operators, thus shaping farmers' behavior by encouraging them to overproduce subsidized crops such as corn and soybeans (Schoonover and Muller 2006), they largely benefit agribusiness firms, food processors, retailers, and restaurant chains that make money by processing these commodities into food products or intermediary food products such as potato chips, soda, fast food hamburgers, corn syrup, vegetable oils, and animal feed (Carolan 2011; Schoonover and Muller 2006). For these companies, keeping commodity prices low through subsidies and overproduction means that the costs of their inputs—corn, soybeans, corn syrup, vegetable oils, livestock fed on inexpensive grains, etc.—stay low, allowing them to gain a greater share of the money consumers spend on food than they otherwise would (Billig and Wallinga 2012; Carolan 2011; Pollan 2003; Tillotson 2004). In other words, due to their political power and ability to influence U.S. farm policy, agribusiness firms and food retailers are able to manipulate the market by keeping the costs of their key inputs artificially low, often below the cost of production, a cost advantage they are then able to take advantage of due to their position in the agricultural commodity chain.

American consumers and most American farmers do not benefit from these policies. Between 1995 and 2009, for instance, approximately 88% of the $211 billion in farm payments made by the U.S. government went to 20% of all U.S. farms, leaving only $24.5 billion over 15 years for the remaining 1.76 million farms (Carolan 2011, 193). Thus, most U.S. farmers receive very little money from U.S. subsidy programs while being stuck in a market in which there is an oversupply of, and thus low prices for, the basic commodities they produce and little government support, and thus more risk, for growing fruits and vegetables (Billig and Wallinga 2012; Schoonover and Muller 2006).

As I discuss in more detail later in the chapter, these subsidies also hurt farmers and food processors in developing nations, who are unable to compete against farmers and agribusiness firms whose sales are subsidized by the U.S. government (Murphy 2009; Pascual and Glipo 2002; Wise 2011; Wise and Rakocy 2010). In addition, by encouraging the overproduction of corn and soybeans and the underproduction of fruits and vegetables, these subsidies hurt U.S. consumers by promoting the production of highly processed, low-nutrition food that costs less at the supermarket than does more nutritious food, thus contributing to a wide range of serious health problems, including obesity, type 2 diabetes, heart disease, high blood pressure, high cholesterol, and stroke (Carolan 2011; Schoonover and Muller 2006).

Production Contracts and Farming Practices

In the U.S. between 2006 and 2008 roughly 99% of broiler (chicken) production occurred under production contracts, approximately 68% of hog production occurred under marketing or production contracts, and broiler and hog producers received about 50% and 19% of the value, respectively, of all production contracts. Moreover, because roughly 20% of all commercial hogs were owned by meatpackers, less than 12% and 1% of hog and broiler sales, respectively, occurred in spot (cash) markets (MacDonald and Korb 2011). The implications of this both at the time and currently are that there is virtually no price discovery in hog and broiler markets,[8] that hog and broiler integrating firms can easily manipulate the contract and spot market prices they pay farmers, and that hog and broiler farmers have little choice but to enter into contracts with large integrating firms (these are the agribusiness companies with which farmers sign contracts). Indeed, in 2005 59% of broiler farmers and 23% of hog farmers had "no marketing option other than their current integrator for the[ir] commodity" (MacDonald and Korb 2008, 16).

As a result, the contractual terms that hog and broiler farmers receive tend to be very onerous. For instance, farmers working under production contracts do not own the hogs or broilers they raise. Instead, the large integrating firms with which farmers sign their contracts own the feed, chicks, hogs, vaccinations, and medicines that the farmers use. These firms provide these inputs to the farmers, pay the farmers to raise the chicks and hogs, determine exactly how the chicks and hogs are to be raised, and provide the farmers with precise design specifications for the facilities used to house the chicks and hogs (Hauter 2009, 28n251). Production farmers, on the other hand, provide the land, facilities, labor, and much of the capital necessary to raise the animals, often going deeply into debt to build or upgrade the facilities in which the animals are housed (Heffernan 2000; James et al. 2012). Production farmers have no say in the production technologies they use (Taylor and Domina 2010), and they are often responsible for handling the waste produced by the chicks or hogs.

Production contracts also tend to cover a much shorter time period than the repayment terms of the loans the farmers take out to build and upgrade their animal production facilities (James et al. 2012). This difference in time schedules is not insignificant since new hog and broiler production facilities cost hundreds of thousands of dollars, facility upgrades can cost tens of thousands of dollars (Hauter 2009), and the median hog and broiler farm has two hog houses and between three and five broiler houses, respectively (MacDonald and Korb 2011). Thus, once farmers enter production contracts with hog or poultry integrating firms, they are dependent on contract renewals to pay back their loans, thereby cementing their dependence on these oligopsony firms.

As the use of contracts has become more widespread and as hog and poultry integrating firms have become increasingly concentrated, the prices farmers receive for their labor have decreased. Research shows, for example, that hog production contracts lower the cash price of hogs sold on spot markets, which, in turn, determines the prices farmers receive for hogs sold under marketing contracts.[9] Moreover, because the use of contracts greatly reduces the number of hogs sold on the spot market, and because daily marketing contract prices are based on the morning price of the spot market, it is very easy for hog integrating firms to keep cash and marketing contract prices low by withholding their purchases until later in the day (Hauter 2009). As a result, the real farm gate price for hogs dropped by 31% between 1989 and 2008, while the real farm gate price for broilers dropped by 26% during this same period (Hauter 2009).

The low prices that hog and broiler farmers receive for their labor, combined with the fact that integrating firms prefer doing business with large farms (Gurian-Sherman 2008),[10] means that the size of hog and broiler farms has increased over time. By 2008, for example, the median "production contract" broiler and hog farm raised 380,000 broilers and 6,000 hogs, respectively, with "half of all [U.S.] hog production occur[ing] on farms that raised at least 15,500 hogs [per year and] half of all [U.S] broiler production occur[ing] on farms that raised at least 682,200 broilers [per year]" (MacDonald and Korb 2011, 18).

In addition to reducing farmer incomes, these large confined animal feeding operations (CAFOs) produce tremendous volumes of animal waste. It is estimated, for instance, that livestock and poultry CAFOs in the U.S. produce 500 million tons of manure per year, more than three times the amount of human feces produced in the nation annually (Pew Commission 2008). This manure, which is geographically concentrated in a few facilities and regions of the country, is generally held, untreated, in large pits or lagoons until it is spread over fields; and nutrients from this waste often end up in the air due to volatilization and in surface and groundwater due to farmland runoff, lagoon and pit leaching, and the periodic accidental spilling of hundreds of thousands and sometimes millions of gallons of waste into local waterways (Gurian-Sherman 2008; Pew Commission 2008).

In turn, these nutrient pollutants can harm terrestrial and aquatic ecosystems, contaminate soil, cause respiratory diseases in humans, and seriously degrade local drinking water quality, possibly causing blue baby syndrome and certain cancers. They can also cause water eutrophication, resulting in massive fish kills, algal blooms, and fecal bacteria contamination (Gurian-Sherman 2008).

CAFOs are also key contributors to the climate change crisis, and they produce noxious odors that reduce nearby property values and severely degrade the quality of life of local residents. Moreover, due to the extremely crowded and unsanitary conditions in which CAFO animals are raised, tremendous amounts of antibiotics, estimated to be eight times the total taken by U.S. citizens for all purposes each year, are used to prevent disease among the confined animals.[11] This, in turn, creates antibiotic resistance among bacteria that cause human disease, resulting in human suffering and reduced economic productivity due to illness (Gurian-Sherman 2008).

Supporters of CAFOs argue that large-scale animal operations are more efficient than are small-scale operations. However, they are only able to make this argument because they ignore the more than $5 billion in costs that the CAFO industry shifts to the public each year in the form of environmental harm, public health problems, reduced property values, and direct and indirect government subsidies (such as subsidies to grain farmers that keep the cost of animal feed artificially low). Once these factors are included in the cost of production, small and midsize farms that use alternative farming methods become much more efficient than large CAFOs (Gurian-Sherman 2008).

It is thus reasonable to conclude that our current meat, egg, and dairy production systems, which all rely on CAFOs and the widespread use of contracts, are driven not by efficiency but by the power of large processing firms that on the one hand play a key role in shaping U.S. agriculture policy and on the other hand use marketing and production contracts, ownership of significant quantities of their own livestock, and oligopsony power in agricultural commodity chains to lock small and medium-sized producers out of the market and shape or dictate the production practices of large producers (Gurian-Sherman 2008). This, of course, is consistent with the IDE argument that elites use organizational, institutional, and network-based mechanisms to monopolize decision making power, shape the behavior of non-elites, and shift social and environmental costs onto others, thereby harming individuals, communities, and the environment.

Captive Value Chains

The type of commodity chain relationship that exists between livestock and poultry integrators on the one hand and contract farmers on the other hand has been described in the literature as a *captive commodity chain relationship*, in which "small suppliers depend on dominant buyers that control and monitor [their] activities" (International Trade Centre 2011, 5). In the previous section I focused on captive commodity chain relationships that exist at a single stage of the commodity chain. However, entire commodity chains, or at least large chunks of these chains, can be characterized as being captive. In such chains, which Gary Gereffi et al. call *captive value chains*, "small suppliers are transactionally dependent on much larger buyers. Suppliers face significant switching costs and are, therefore, 'captive.' Such [value chain] networks are frequently characterized by a high degree of monitoring and control by lead firms" (2005, 84).

In identifying, defining, and empirically confirming the existence of this and other types of commodity chain, Gereffi et al. demonstrate that firms do not have to directly own the various stages of production in the chain to control and coordinate the activities of other commodity chain participants. Instead, the dominant firms in the chain, generally referred to as lead firms, can control and coordinate these activities by exerting their oligopsony or monopsony power down the chain, often benefiting from the ability of their first-tier suppliers to, in turn, exert oligopsony power over suppliers located even further down the chain.

In the food and agriculture industry, large supermarket retailers such as Walmart, Kroger, and Tesco often take on the role of lead firm in what can best be described as captive agricultural value chains, with devastating results for small and midsize farms around the world. An important example of this is provided by Catherine Dolan and John Humphrey (2001, 2004), who document the development, in the 1990s, of a set of captive value chains linking supermarkets in the United Kingdom to exporters and farmers in Kenya. They argue that prior to the development of these captive value chains, it was relatively easy for Kenyan farmers and produce exporters to access UK produce markets. In the 1980s, for example, over 100 Kenyan companies exported produce that they had purchased on rural spot markets from an estimated 15,000 Kenyan smallholder farmers who, in turn, grew nearly three-quarters of Kenya's fruits and vegetables.

This situation changed drastically in the 1990s as supermarkets in the UK began to (a) capture an ever larger share of that nation's fresh fruit and vegetable sales and (b) bypass British wholesale markets in favor of working directly with a handful of British importers. Bypassing wholesale markets allowed British supermarkets to gain greater control over the quality, specifications, and production standards of the fruits and vegetables they sold. It also allowed them to "delegat[e] lower-profit functions such as quality control, monitoring, and distribution to [their] importers" (Dolan and Humphrey 2004, 497).

The shift away from wholesale markets also resulted in each supermarket chain restricting the number of produce suppliers (importers) it sourced from. In turn, each supplier now tended to work with a single exporter in each African nation, while each African exporter worked with a single British supplier. This, of course, made the suppliers extremely dependent on specific supermarket chains and the exporters extremely dependent on specific suppliers. This dependence, in turn, meant that African exporters and farmers increasingly had to tailor their produce and production methods to the

> product and process parameters of UK supermarkets, . . . forcing them to acquire a range of new capabilities to retain their UK business. They could no longer act purely as growers or traders. Product and process innovations depended on sophisticated technical knowledge of production as well as on close ties with researchers, seed companies, and importers. The emergence of semiprocessed ready-to-eat products, for instance, entailed heavy investments in cold storage, packhouses, and high-care facilities, so that produce could be harvested, processed, and transported to the United Kingdom in hygienically and temperature-controlled conditions. . . . By the end of the 1990s the[se] [increased] demands for capital and technical capacity . . . led to the exclusion of many small exporters that were unable to meet supermarket requirements. This exclusion was clearly evident in all the major African fresh fruit and vegetable exporting countries but was particularly significant in Kenya, where the top-seven firms controlled over 75% of all exports by the end of the 1990s. (Dolan and Humphrey 2004, 501)

In addition to drastically reducing the number of Kenyan produce exporters and forcing these exporters to tailor their production methods to the demands of British supermarkets, the development of these captive value chains led exporters to source produce almost entirely from large farms, thus cutting thousands of small farms out of the market.

As a result, small farms' share of Kenya's export market declined from just under 75% in the early 1990s to 18% in 1998 (Dolan and Humphrey 2004). Because exporters were forced to cover many of the financial costs associated with upgrading their production, transport, and storage technologies, their profit margins also went down, resulting not only in increased export volume (Dolan and Humphrey 2004) but also in the passing of a significant portion of the new costs onto growers (International Trade Centre 2011), a finding confirmed for captive agricultural commodity chains examined in other studies as well (Farquhar and Smith 2006; Tallontire and Vorley 2005).

Indeed, a survey of the literature on agricultural commodity chains that summarized the findings of 37 peer-reviewed journal articles, 22 research institution reports, and four books or book chapters that were published between 2000 and 2011 (International Trade Centre 2011) found that the vast majority of research on the topic uncovered findings similar to those reported by Dolan and Humphrey. Covering products as diverse as coffee, tea, cotton, flowers, fish, fresh fruits, vegetables, and cocoa, and global commodity chains as well as chains centered in Europe, Africa, Latin America, and Asia, the survey found that the emergence of captive agricultural value chains in which very specific product and process standards are imposed on the chain by lead firms has made market participation by smallholder farmers around the world exceedingly difficult (due to very high barriers to entry) and thus much less likely to occur.

The rise of these retailer-led captive agricultural value chains has also likely harmed the environment. It is true, of course, that the ability of supermarkets to determine what crops are grown, what seeds are used, what pesticides and fertilizers are applied, and how they are applied means that supermarkets may sometimes be able to improve the environmental conditions under which agricultural production occurs (International Trade Centre 2011). However, the more significant trend in captive agricultural value chains is that they shift agricultural production from smallholder farms to large farms that due to their size require the use of environmentally damaging modern agricultural techniques (small farms often use these techniques, but large farms have to do so). Moreover, in order to satisfy British supermarkets, large Kenyan exporters had to work closely with researchers and seed companies (Dolan and Humphrey 2004), which is potentially quite problematic since (a) most seed research produces seeds that require the intensive use of synthetic fertilizers and pesticides, (b) the world's largest seed companies manufacture seeds that generally require these same inputs, and (c) several of the world's largest seed companies are also among the world's largest agrochemical firms and thus make immense profits from selling environmentally harmful pesticides and fertilizers.

To sum up, one of the key arguments in this and the preceding section is that large agribusiness firms and retailers do not have to directly own specific stages of the agricultural commodity chain to strongly influence, and in many cases dictate, the farming practices and production methods carried out by the actors operating in these stages. Large agribusiness firms and retailers thus play a key role in ensuring that farmers around the world adopt and use harmful agricultural technologies rather than alternative technologies that are much less socially and environmentally harmful. This should not be surprising, of course, since many of these companies directly benefit from the use of socially and environmentally degrading agricultural practices, either because these practices reduce the costs of their inputs or because they own or have formed partnerships with companies that provide the technologies and equipment that undergird these practices.

Another key point in this and the preceding section is that oligopsony firms in the food and agriculture industry use their buyer power to distort agricultural markets, unfairly lower the prices that farmers receive for their goods, and marginalize small and medium-sized farms in developing *and* developed nations. On the one hand, this demonstrates that agricultural markets do not in any way resemble the ideal of free and competitive markets that many people argue undergird the world economy. On the other hand, it demonstrates that farmers around the world are similarly disadvantaged by an agricultural system that treats them all as cogs in a machine. Finally, it shows that large agribusiness firms and retailers use a set of mechanisms derived from their oligopsony power in agricultural commodity chains to accumulate capital, monopolize decision making power, shape farmer and supplier behavior, and shift the social, financial, and environmental costs of their activities onto others. Since commodity chains are organizational networks, and since oligopsony power is an organizational and network-based phenomenon, these findings are consistent with my theoretical argument.

The Seed Market

As is true of firms that exercise buyer power in agricultural commodity chains, agribusiness firms that sell seeds, agrochemicals, and other agricultural inputs are also highly concentrated, allowing them to exert oligopoly power over farmers. In order to better understand how oligopoly power works in agricultural commodity chains, this section briefly examines the seed industry, with a particular emphasis on Monsanto, the largest seed company in the world.

In 2007 Monsanto controlled 23% of the global proprietary seed market, followed by DuPont at 15%, Syngetna at 9%, and Limagrain at 6% (ETC Group 2008).[12] This high level of corporate concentration is paralleled in the U.S., where a small handful of firms, led by Monsanto, also dominate the seed market. Monsanto, for instance, directly controlled 36.5% of U.S. corn seed sales in 2008, followed by DuPont/Pioneer at 30% and Syngenta at 10%. Moreover, due to licensing agreements with over 200 other seed companies, Monsanto indirectly controlled an additional 23.5% of the U.S. corn seed market, with corn seed containing genetic material owned by Monsanto planted on 80% of the cornfield acreage in the U.S. In 2008 Monsanto also directly controlled about 30% of the U.S. soybean seed market, a figure that rises to over 60% of the market and 91% of the acreage when licensing agreements with other firms are included in the tally (Hubbard 2009).

The high level of concentration that exists in the U.S. crop seed industry results largely from changes in U.S. patent law since 1970 that have made it profitable for large chemical and pharmaceutical firms to acquire, merge, or partner with their seed company competitors. Prior to 1970, U.S. patent law prevented the patenting of sexually reproducing plants and their seeds. This changed in 1970, with the passage of the Plant Variety Protection Act (PVPA), which provided 20 years of patent protection for most crops and seed. Congress intentionally included two exceptions in the law: researchers outside the patent-holding company could use the germplasm from patented seeds to conduct plant and seed research, and farmers could save patented seed from their harvested crops for replanting. This changed in 1985 when the U.S. Patent and Trademark Office (U.S. PTO), reacting to a 1980 Supreme Court decision, decided to allow companies to patent sexually reproducing plants under patent utility law rather than the PVPA. In doing this, the U.S. PTO not only appeared to contradict the will of Congress, which had explicitly chosen not to include sexually reproducing plants under patent utility law, it also failed to provide the patent exemptions that Congress had previously established for researchers and farmers. The U.S. PTO's decision, which was upheld by the Supreme Court in 2001, thus greatly increased the ownership rights granted to seed patent holders while completely curtailing the rights of non–patent holders to use patented seeds without the patent holder's permission (Center for Food Safety 2005; Hubbard 2009).

In addition to curtailing the rights of non–patent holders, these changes in U.S. patent law gave corporations the power to patent specific genetic sequences, the seeds and plants derived from these genetic sequences, engineering techniques used to create genetically modified seeds and plants, and genetic mechanisms found in the natural world. As a result, any company, researcher, or farmer that wants to use a genetically engineered seed trait or technique that is patented by someone else must sign a licensing agreement with the patent holder, who, in turn, can decide to whom and under what terms to grant licenses. This makes it easier for the patent holder to shape farmer behavior, reduce market competition, and restrict seed and plant research carried out by others. It also makes it easier for patent holders to generate profits from their patented seed (Center for Food Safety 2005; Hubbard 2009).

Armed with these patent protections, large chemical and pharmaceutical firms, which were well positioned to enter the genetically engineered seed business, began to acquire or join forces with their seed industry competitors. Between 1996 and 2008, for example, Monsanto, DuPont, Syngenta, Bayer, Dow, and BASF collectively acquired or formed joint ventures with over 200 seed and biotech companies, with Monsanto acquiring more than 50 seed and biotech firms, including some of the world's largest, at this time (Howard 2009). During this period Monsanto also formed joint ventures with Cargill, Dole, BASF, and Dow (ETC Group 2008; Howard 2009) and signed cross-licensing agreements with some of its largest competitors, including DuPont, Syngenta, Bayer, Dow, and BASF (Dow had cross-licensing agreements with all but one of these companies; Howard 2009). And finally, as part of this process, more than 200 independent U.S. seed companies went out of business or were acquired by the industry giants (Hubbard 2009, 4).

Large seed companies such as Monsanto made these acquisitions and entered into these joint ventures not only to gain market share and eliminate direct market competition but just as importantly to gain ownership of genetically engineered traits, genetic engineering techniques, and germplasm that they would then control.[13] This was important to them because it eliminated the fees they would otherwise have had to pay for licensing agreements; removed genetic traits, techniques, and germplasm from the hands of their competitors; and allowed them to license a wider array of traits to other seed companies, thereby giving them the power to reduce market competition, shape farmer and competitor behavior, and drastically increase the prices of their seeds.

Monsanto has used this power very effectively. For example, a lawsuit filed in 2007 claims that in the late 1990s Monsanto used licensing agreements with small seed companies that wanted to use its proprietary technology to force these companies to include Monsanto traits in at least 70%–85% of the genetically engineered seed they sold. The lawsuit further alleges that these agreements, some of which lasted for 10 years, prevented licensees from combining, or stacking, Monsanto-derived seed traits with seed traits created by other companies, thereby reducing competition and innovation in the industry and making it more difficult for farmers to access seed traits developed by Monsanto's competitors (Hubbard 2009). Monsanto has also been accused of entering into agreements with other seed trait developers that restrict the ability of these developers to "market [their] trait[s] outside the agreement" (Moss 2011, 98).

The high level of control that Monsanto and other large seed companies exert over the U.S. seed market, combined with the fact that 90% of U.S. corn and upland cotton acreage and 93% of U.S. soybean acreage is planted with genetically engineered seed (USDA 2013), most of which is produced by Monsanto, has also meant that Monsanto and other major seed companies have been able to drastically increase the prices they charge farmers for their seed. Between 1999 and 2009, for example, the "prices farmers paid for seed in the U.S. increased by 146%, with 64% of that increase occurring in the last three years of that period" (Hubbard 2009, 21). In addition, the price of Roundup herbicide, which has to be used by farmers who plant Monsanto's herbicide-resistant Roundup Ready seeds, nearly doubled between 2006 and 2009 (Hauter 2009).

According to Kristina Hubbard, director of advocacy and communications at the Organic Seed Alliance,

> Biotechnology traits and [associated] technology fees are the driving force behind increased seed costs. . . . For example, [Monsanto's] Roundup Ready trait in soybeans [which] added $6.50 per bag in 2000 . . . now cost[s] $17.50 per bag. . . . This means a farmer who plants one bag of Roundup Ready soybeans per acre on 1,000 acres has seen his production costs increase by $11,000 . . . due to the trait price increase alone. It also means that smaller seed companies that license the trait for varieties they have developed independently recoup only a fraction of their research costs, since much of the price goes back to Monsanto in the form of a royalty. (2009, 22)

Farmers are forced to pay these high prices for several reasons. First, seed companies impose royalty fees for *each* genetically modified trait found in a seed regardless of whether the farmer wants or needs each trait. Second, due to the oligopoly power of the large seed companies, many farmers are finding it nearly impossible to obtain high quality non–genetically modified (conventional) seed, which is apparently unavailable in some regions of the country, and increasingly difficult to find seeds that have fewer than two or three genetically modified traits (Benbrook 2009; Hubbard 2009). Third, farmers have to sign technology agreements when they purchase genetically engineered seed. These agreements generally prevent farmers from saving the seed they just bought, from using it in any but one season's crop, and from replanting or giving away harvested seed, thus forcing them to buy new seed every year.

These technology agreements also require farmers to resolve disputes with the seed company "through binding arbitration or in a court convenient to the company" and to let "seed company representatives inspect their fields" to ensure that the agreements have not been violated (Hayes 2009, 9). Farmers are expected to follow strict farming practices, they cannot give the modified seed to any other entity, and because genetically modified crops are not accepted in all agricultural markets, they must agree to keep their genetically modified crops out of these markets (Hayes 2009).

Farmers who violate these technology agreements can be sued or forced to settle out of court for patent infringement by the seed company. For instance, as the most aggressive of the large seed companies, Monsanto investigates hundreds of farmers for such infringements each year, resulting in 112 patent infringement lawsuits and hundreds of out-of-court settlements against farmers as of late 2007, for which the company had won court damages of over $21 million and received out-of-court settlement fees of between $85 million and $161 million (Center for Food Safety 2007).

Moreover, farmers who have not knowingly purchased or planted genetically modified seed from Monsanto in a particular year have also been successfully sued for patent infringement by the company after it was determined that Monsanto's genetic material had made its way into their fields that year: this has apparently occurred even when seed legally planted one year accidentally germinated in the field the following year, when seed from the previous year's crop accidentally fell from the plants onto the field prior to harvest, and when the farmer's fields were genetically contaminated due to pollen drifting from neighboring fields or from seed blowing off passing trucks (Center for Food Safety 2005).

Despite these problems, many farmers still want to plant genetically modified seed because they have been told that using genetically modified seed will increase their crop yields and reduce their use of herbicides and insecticides (Benbrook 2009; Gurian-Sherman 2009), thereby increasing their profits. Research demonstrates, however, that genetically modified seeds do not increase yields relative to conventional seeds or organic farming practices (Gurian-Sherman 2009).[14] Nor does the use of genetically modified seed reduce pesticide use. In fact, between 1996 and 2008 the adoption of genetically engineered corn, soybean, and cotton resulted in the application of 318.4 million *additional* pounds of pesticides in the U.S. and the emergence of glyphosate-resistant weeds that now infest millions of acres of U.S. farmland (glysophate was the most used herbicide in the U.S. in 2007; Benbrook 2009).

The widespread adoption of genetically engineered seeds and crops has thus had several negative social and environmental consequences, including (a) the loss of hundreds of small, regional seed companies that were more likely than the international giants to produce seeds bred for local conditions (Hubbard 2009); (b) greater insecurity for farmers as the prices of their inputs increased dramatically (Carolan 2011; Hubbard 2009) and as they became increasingly concerned about being sued by Monsanto (Center for Food Security 2005; Hayes 2009; Hubbard 2009); (c) a loss of farmers' control, or choice, over how they farm their land (Hauter 2009; Hayes 2009); and (d) a greater reliance on toxic pesticides than would otherwise be the case. Genetically altered material also appears to be spreading from genetically modified crops to conventional crops and wild species, including weeds (Benbrook 2009; Center for Food Safety 2005; Dunfield and Germida 2004; Eastham and Sweet 2002; Gurian-Sherman 2009; Oliveira-Souza 2000), a development with unknown but potentially dire consequences.

The oligopolization of the seed industry and its reliance on genetic engineering, highly restrictive patents, and a small number of crops and crop varieties also contributes to the long-standing decline in crop genetic diversity that I discussed earlier, making crops more susceptible to pests, disease, and climate change (De Schutter 2009; Howard 2009; Hubbard 2009). The increasing use of genetically engineered seeds and restrictive patents also reinforces the world's reliance on environmentally harmful agricultural practices by (a) making it more difficult for farmers to purchase seed that does not require heavy pesticide and fertilizer use, (b) reducing farmers' profit margins so that they are forced to cultivate ever larger plots of land, which can only be done using environmentally destructive farming techniques, and (c) greatly enhancing the commodity chain power of agribusiness corporations that directly profit from the use of environmentally and socially damaging agricultural technologies.

It is quite clear, then, that oligopoly power in agricultural commodity chains is a key mechanism used by large seed companies to accumulate capital and shift the social, economic, and environmental costs of their emerging technologies onto others.

Commodity Chains, Farmers, and the Environment

At the beginning of the chapter, I argued that oligopoly and oligopsony firms are able to undercut market competition, shape the behavior of other market actors, and distort markets and prices to their benefit by forcing buyers and sellers to accept terms of exchange dictated by the oligopoly or oligopsony firm, by using information that they uniquely possess (because of their size, resources, and oligopoly or oligopsony position) in ways that disadvantage buyers, sellers, and competitors who lack this information, by flooding markets or selling goods at below cost, and by using the financial resources they generate in distorted markets to influence government policies in ways that produce ever more favorable market outcomes for them.

I then demonstrated that oligopoly and oligopsony firms do, in fact, use the first two of these four mechanisms to undercut market competition, shape behavior, and distort markets and prices. Specifically, I explained how large agribusiness firms use production and marketing contracts, commodity futures markets, and the resulting division of agricultural market exchanges across multiple cash and contract markets to distort prices to their benefit. I further explained how the ability of these firms to distort these "divided" markets is based largely on the informational advantage they possess vis-à-vis farmers regarding what is happening in these multiple markets, on the fact that these divided markets are smaller than the overall market, making it easier to manipulate each of them, and on the fact that outcomes in each of these markets influences, to some degree, outcomes in the other markets.

I also demonstrated that the lack of a viable cash market for most livestock producers and the inadequate market supply of conventional corn, soybean, and cotton seed lowers the prices that livestock producers receive for their labor, increases the prices that farmers pay for their seed, and forces both groups to sign contracts that on the one hand dictate the materials and production methods they use[15] and on the other hand leave them vulnerable to additional corporate abuses. Because these commodity chain dynamics and those discussed in the previous paragraph all push down farmers' profit margins, and because large agribusiness firms prefer working with large suppliers, the increasing oligopsony and oligopsony power of agribusiness firms also encourages the growth of large farms while simultaneously pushing many small and midsize farms out of the market.

In this regard, it is important note that though we normally do not consider farmers and agribusiness firms to be competitors, the fact is

that they do compete to a significant degree for the same set of consumer dollars, such that actions that increase agribusiness profits often
come at the expense of farmers, whose share of each consumer dollar
spent on food declined from 50 cents in the late 1940s to 19 cents in
2006, with most of this decline captured by agribusiness firms, supermarkets, and other corporate actors (Carolan 2011; Domina and Taylor
2010). It is thus clear that the commodity chain dynamics highlighted in
this chapter are used by agribusiness firms and supermarkets to undercut farmer competition for the money that consumers spend on food.

Finally, I demonstrated that in many parts of the world, the captive
value chain relationships that exist between supermarkets and produce
suppliers, exporters, and farmers provide supermarkets with the power
to dictate which fruits and vegetables farmers grow, which seeds farmers
use, and which production technologies farmers, exporters, and suppliers adopt. Because supermarkets use their oligopsony power to push the
costs of production down the commodity chain and because exporters
prefer to work with large farms, the creation of supermarket-led captive value chains has also reduced exporters' and farmers' profit margins
while simultaneously pushing many small and midsize farms out of the
supply chain, encouraging yet again the growth of large farms.

The result of all this in the U.S. and around the world is declining
farmer incomes, the increasing marginalization of millions of small and
midsize farms, the continued adoption of socially and environmentally
harmful agricultural production practices, and the ever-growing power
of agribusiness firms, which profit immensely from the harm they do to
the environment and others.

Further enhancing the power of agribusiness firms are a set of organizational and institutional mechanisms I have not yet discussed, including international free trade agreements that facilitate the global spread
of agricultural commodity chains and international property rights
treaties that help cement corporate control over these chains by extend-
ing to nearly all the nations in the world the type of patent protections
that seed companies receive in the U.S. Thus, in the remaining sections
of this chapter, I discuss one of the world's most important free trade
agreements, the World Trade Organization's Agreement on Agriculture,
and one of the world's most important property rights agreements, the
WTO's Agreement on Trade-Related Aspects of Intellectual Property

Rights. In addition to explaining how these agreements contribute to the outcomes listed in the preceding paragraph, I use my case studies of these two agreements to demonstrate that agribusiness firms undercut their competition and distort agricultural markets by (a) flooding these markets and selling goods at below cost and (b) using their political power to create international treaties that produce increasingly favorable market outcomes for them. These case studies thus provide strong support for my earlier claims that agribusiness firms use free trade agreements, international property rights protections, and their ability to influence government policy to achieve their capital accumulation goals.

Liberalizing Agricultural Trade

The goal of free trade agreements is to liberalize trade, which refers to the easing of restrictions and elimination of policies that make trade between nations difficult. In liberalizing trade, these agreements thus increase trade between nations and provide firms with new markets, new investment opportunities, and increased access to agricultural products and natural resources in nations around the world. Important policies and restrictions targeted by free trade agreements include import and export quotas, tariffs, preferential government purchases of locally made products, technical rules favoring goods produced by specific nations, and government subsidies that provide a nation's producers with an unfair advantage in international markets by allowing them to sell goods at below cost.

As noted in chapter 3, trade liberalization is one of the key goals and key consequences of World Bank and IMF structural adjustment policy. It has also been achieved through the ratification of international free trade agreements such as the North American Free Trade Agreement (NAFTA), which covers the U.S., Mexico, and Canada, and the set of agreements that created the World Trade Organization (WTO), which includes as members most of the nations of the world.

Proponents of trade liberalization argue that free trade is good for all parties because it allows nations to export goods that they produce most efficiently while importing goods that they produce less efficiently, thus benefiting all nations and citizens. The reality, of course, is that trade liberalization does not benefit all parties, and even among those that do benefit, some benefit more than others.

Agricultural trade liberalization, a centerpiece of many structural adjustment loans and free trade agreements, is a case in point. Rather than aiding all parties, agricultural trade liberalization tends to increase agribusiness and supermarket profits while simultaneously harming farmers and the environment around the world. It does this in several ways. First, as agricultural trade becomes more liberalized, agribusiness corporations and food retailers have access to a greater number of farmers and nations that they can play off against each other (Wallach and Woodall 2004), thereby ensuring that these corporations can source from the lowest cost suppliers possible. Second, more farmers competing in international markets means more production devoted to international trade (increased supply), which lowers the prices that agribusiness firms and retailers pay farmers, pushing farmers to increase production, thereby lowering prices further (Vorley 2004). Third, the opening of multiple national markets means that agribusiness corporations can sell agricultural products below cost in one nation, often by flooding that nation's markets with low-priced goods, and make up for their losses elsewhere, thereby putting farmers and smaller competitors in the targeted nation out of business and increasing the share of the national and global market that the large agribusiness firms control (Heffernan 2000). Fourth, because wealthy nations such as the U.S. tend to provide large subsidies to their farmers even after signing international free trade agreements, large agribusiness firms—whose input costs are significantly lower due to these subsidies—can sell agricultural goods in developing nations at prices that are often below the cost of production in these nations, thereby undermining the principles of fair trade that are supposed to underlie free trade. This puts local farmers and food processors around the world out of business, not necessarily because they produce agricultural goods inefficiently but simply because their governments cannot subsidize agricultural production (Murphy 2009).[16]

Wise and Rakocy (2010) demonstrate, for instance, that the creation of NAFTA in 1994 allowed U.S. agribusiness firms to more easily export hogs raised on feed made from subsidized corn and soybeans. As a result, U.S. pork exports to Mexico increased by 700% from the early 1990s to 2008, while Mexican pork prices dropped by about 56% during roughly the same time period, throwing many small and medium-sized Mexican hog producers out of business. At the same time, greatly increased U.S. exports of subsidized corn and soybeans to Mexico (sold on average at 19% and 12% below the cost of U.S. production, respectively, between 1997 and 2005) played a key role in devastating Mexico's agricultural sector, with roughly 2.3 million Mexicans leaving the sector between 1993 and 2008.

Not surprisingly, and for similar reasons, the creation of the WTO in 1995, with its concomitant increase in subsidized agricultural exports, weakened the viability of small farms around the world (ICTSD 2009), resulting in hunger and widespread loss of farmers' livelihoods (Murphy 2009; National Academy of Agricultural Sciences 2006; Pascual and Glipo 2002; Wallach and Woodall 2004). Between 1997 and 2003, for instance, U.S. wheat, soybean, corn, cotton, and rice exports were sold on average at prices 37%, 11.8%, 19.2%, 48.4%, and 19.2% below the cost of production, respectively, making it very difficult for developing nation farmers to compete in local and international markets (Murphy et al. 2005). Moreover, wealthy nations continued subsidizing agricultural production after 2003 by shifting subsidies from categories that are banned by the WTO to categories that are still allowed by the WTO despite the fact that these subsidies "distort trade, harm the environment, and hurt developing nation farmers" (ICTSD 2009, 1).

High levels of agricultural trade liberalization can also make it difficult for developing nations to feed themselves. In Guatemala, for instance, trade liberalization and the development of export-oriented agriculture in the 1990s and the first few years of the 21st century led to a 170% increase in staple commodity imports and to a decline in agricultural production for local consumption, such that by 2006 only 20% of the food consumed in the country was produced in the country (Ziegler 2006); and in Haiti trade liberalization led to the failure of the country's previously self-sufficient rice market and to such heavy dependence on rice imports that by 2011 between 75% and 80% of the rice consumed in Haiti was imported (Curran 2012; Suarez and Rubio 2011). Under such circumstances, dramatic increases in global food prices can lead to severe hunger and food riots, as happened in Haiti in 2008 (Quigley 2008), highlighting the vulnerability to price swings and food supply disruptions that countries such as Haiti and Guatemala face as a result of agricultural trade liberalization.

Because liberalized agricultural trade reduces farmers' profit margins (by reducing the prices farmers receive for their goods) and undermines the ability of small farms in developing nations to survive, it also increases the number of large farms in these nations, farms that are more likely to produce for export markets, to use mechanized and chemical intensive farming practices that cause severe environmental damage, and to rely on a narrow genetic pool for their crops (Kennedy et al. 2004; Nadal and Wise 2004; Wallach and Woodall 2004). Liberalized agricultural trade also harms the environment by increasing the volume of international trade, which requires vast quantities of fuel (Wallach and Woodall 2004), and by increasing poverty and landlessness in developing nations, which tend to increase these nations' rates of deforestation and soil erosion (Bello et al. 1999; Boyce 2002). As a result, liberalized agricultural trade and the mechanisms that make liberalized agricultural trade possible, such as free trade agreements and structural adjustment, simultaneously promote environmental and social degradation. They also promote capital accumulation by providing agribusiness firms with new markets around the world and by allowing these firms and food retailers to source their agricultural inputs from as many nations and farmers as possible, thereby increasing the volume and lowering the prices of these goods on the world market.

Thus, in the following section, I examine the organizational, institutional, and network-based mechanisms that economic and political elites in the U.S. and Europe used to draft and ratify the World Trade Organization's Agreement on Agriculture (in itself an important elite-controlled mechanism), focusing in particular on the role that members of the U.S. power elite and policy planning network played in formulating U.S. agricultural trade policy in the 1970s and on the decision making rules that provided U.S. and European officials with the power to achieve their trade goals during the negotiations that produced the Agreement on Agriculture.

The WTO's Agreement on Agriculture

The WTO and the Agreement on Agriculture (AoA) were created during the Uruguay Round of GATT negotiations that took place from 1986 to 1994. GATT, which stands for the General Agreement on Tariffs and Trade, was created shortly after World War II (1947) as a forum for nations to negotiate tariff reductions so as to increase global trade and prevent another global economic depression. The creation of the WTO greatly extended the scope and reach of GATT beyond what had been attained in previous negotiating rounds, in large part through the ratification of agreements such as the AoA.

The AoA, which came into effect in 1995, dramatically reduced agricultural trade barriers between most of the world's nations (Wallach and Woodall 2004).[17] The original draft proposal for the AoA was supposedly written by Dan Amstutz, a former high-level executive in the agribusiness firm Cargill, who was ambassador and chief U.S. negotiator for agriculture during the Uruguay Round negotiations from 1987 to 1989 (Murphy 2002; Wallach and Woodall 2004).[18] Regardless of whether Amstutz actually wrote the original draft proposal, the AoA was the product of long term thinking about U.S. economic growth and liberalized agricultural trade that dates back to at least the early 1970s, when representatives of agribusiness and the U.S. power elite convinced the Nixon administration to push for liberalized agricultural trade.

The story begins in 1970, when President Nixon, worried about severe structural weaknesses in the U.S. economy, created an executive branch advisory committee known as the Williams Commission to advise him on international economic policy.[19] In July 1971 the commission released a report that argued that the U.S. economy had great comparative advantage in two areas, high-tech manufactured goods and agriculture (Williams 1971). The commission further argued that the U.S. should work strenuously to open up agricultural markets in Europe and the developing world and that this goal should be accomplished largely through GATT negotiations directed toward abolishing the agricultural trade distorting practices of the European Community and developing world. Citing the Williams Commission report, Peter G. Peterson, assistant to President Nixon for international economic affairs, made the same basic argument in a December 1971 report to the president and Congress (Peterson 1971); and in its 1973 and 1974 reports to the president and Congress, Nixon's Council on International Economic Policy (CIEP), headed by Peter Flanigan, also argued forcefully for liberalizing agricultural trade through GATT negotiations (Flanigan 1973b, 1974).

The Nixon administration took the recommendations of the Williams Commission seriously and quickly adopted a policy of liberalizing agricultural trade through GATT negotiations (Krebs 1999), a position the U.S. had not previously taken (Barkema et al. 1989). For instance, in September 1971, two months after the release of the Williams Commission report, "the Department of Agriculture explained that 'a major consideration in Nixon's New Economic Policy (NEP) [was] the need for American agriculture to remain a growth factor and to continue expanding its markets abroad'" (Burbach and Flynn 1980, 49). In addition, in 1972 Peter Flanigan, Peterson's successor as assistant to President Nixon for international economic affairs, "implor[ed] the [U.S. Department of Agriculture] to develop a strategy for the upcoming GATT negotiations" (Krebs 1999, 5).

In 1972 the Nixon administration also appointed William Pearce, a former vice president of the agribusiness firm Cargill and "a leading architect of the Williams commission report," as a special deputy trade representative for the U.S. (Burbach and Flynn 1980, 249). Pearce also played a leading role in getting Congress to pass the Trade Reform Act of 1973, which "directed U.S. [trade] negotiators to trade off concessions from the U.S. in the industrial sector in exchange for concessions to the U.S. in the agricultural sector" (Krebs 1999, 5). Finally, in 1973 Peter Flanigan submitted a report to Congress outlining the best way to incorporate agriculture in the upcoming Tokyo Round GATT negotiations (Flanigan 1973a). It is quite clear, then, that by 1973 the Nixon administration had made liberalized agricultural trade an important component of its trade policy and one of its key negotiating items in the upcoming Tokyo Round of GATT negotiations, which lasted from 1973 to 1979.

Despite efforts by both the Nixon and Carter administrations to liberalize agricultural trade through GATT, the Carter administration eventually decided that this was not an achievable goal. As a result, U.S. efforts to liberalize agricultural trade made few gains during the Tokyo Round (Burbach and Flynn 1980). However, the Reagan administration decided as early as 1981 to make liberalized agricultural trade a key component of the next round of GATT negotiations (Croome 1999, 6), a decision that was later supported by Presidents Bush and Clinton and that eventually resulted in the adoption of the AoA in 1995.

Although the Reagan administration took a much stronger and more aggressive position on agricultural trade during the Uruguay Round of GATT negotiations than the Nixon and Carter administrations took during the Tokyo Round, the key point is that every president since

Nixon has supported liberalizing agricultural trade through GATT negotiations.[20] Therefore, Peter Flanigan, Peter Peterson, and the members of the Williams Commission played key roles in shaping what is now more than 40 years of U.S. agricultural trade policy. In the next section, I examine who these key players were and ask whether they were members of the power elite or policy planning network, as power structure researchers suggest should be the case (see chapter 2 for a summary of power structure theory).

The Social and Professional Backgrounds of the Key Players

The Williams Commission was headed by Albert L. Williams, chairman of the finance committee and former president of IBM. In 1970, the year the commission was formed, Williams sat on the corporate boards of Mobil Corporation, Citibank, Eli Lilly, and General Foods and prior to 1970 he also sat on the corporate boards of General Motors, Allied Chemical, J. Henry Schroder Banking Corporation, and Intertype Corporation. Two members of the Williams Commission, William Pearce and Edmund Littlefield, played especially important roles in promoting the importance of agriculture in the commission (Burbach and Flynn 1980; Krebs 1999). In 1970 Pearce was vice president of public affairs for Cargill; from 1971 to 1974 he was U.S. deputy special trade representative with the rank of ambassador; and as noted earlier, he helped guide important trade legislation through Congress in 1973. Pearce later sat on the boards of several prominent investment firms and joined the Trilateral Commission in 1977.

In 1970 Edmund Littlefield was the president and general manager of Utah Construction and Mining Company, which was soon to merge with General Electric, eventually making Littlefield one of the wealthiest people in the U.S. In 1971 he served on the board of directors of General Electric, Industrial Indemnity Company, Marcona Corporation and Subsidiaries, Pima Mining Company, Hewlett-Packard, First Security Corporation, Wells Fargo Bank, Chrysler Corporation, and Del Monte; and from 1975 to 1976 he was the chairman of the prestigious Business Council, which meets regularly with government officials and is one of the most important organizational links tying the power elite to the U.S. government (Domhoff 2002).

Thus, in 1970 Williams and Littlefield were both members of the power elite as defined by G. William Domhoff (Pearce does not appear to have been a member of the power elite in 1970, but he was an agriculture industry executive). Peter Flanigan and Peter Peterson were also members of the power elite in 1970. Flanigan was born into the upper class. His father, Horace Flanigan, was chairman of Manufacturers Hanover Trust, and his mother was an heiress to the Anheuser-Busch family fortune. In 1970 Flanigan was vice president of Dillon, Read and Company, a major Wall Street investment firm, and before joining the Nixon administration in 1969, he was on the board of directors of Adolphus Busch Estates Inc. and United Gas Corporation. He also later became the director of the U.S. Chamber of Commerce and a trustee of the conservative John M. Olin Foundation.

Peter Peterson became president and CEO of Bell and Howell Company in the early 1960s, a position he held until he joined the Nixon administration in 1971. In 1970 he sat on the corporate boards of First National Bank of Chicago, Illinois Bell Telephone, and Diebold Venture Capital Corporation. He was also a trustee of the Committee for Economic Development, the Brookings Institution, and the University of Chicago. He joined the Trilateral Commission in 1973 and in that year was also a board member of American Express, General Foods, 3M Company, and RCA. In the 1970s he also sat on the board of directors of the Council on Foreign Relations, one of the most important foreign policy think tanks in the U.S.

Thus, the agricultural trade policies adopted by the Nixon administration in the early 1970s were conveyed to the U.S. government by representatives of the U.S. power elite and U.S. agribusiness, both as members of an important nongovernmental advisory committee (the Williams Commission) and as officials and trade representatives within the Nixon administration. These policies were developed in a context in which members of the power elite and the corporate and agribusiness community defined the causes of and general solutions for the economic crisis facing the U.S., and the principal aim of these policies—reducing barriers to agricultural trade through GATT negotiations—has remained an important policy goal of every president since Nixon.

High level officials from the agriculture industry also played key roles in shaping the Tokyo and Uruguay Round GATT negotiations. For instance, former Cargill vice president William Pearce served as deputy special representative for trade negotiations in the Nixon administration, former Cargill vice president Dan Amstutz was the chief U.S. negotiator for agriculture during the Uruguay Round negotiations from 1987 to 1989, and former member of the ConAgra board of directors and Chicago Mercantile Exchange president and CEO Clayton Yeutter headed the U.S. Trade Representative's Office from 1985 to 1989. In addition, the president and CEO of Cargill "was on the GATT Advisory Committee to the US Government throughout the Reagan, Bush, and Clinton Administrations" (Desmarais 2003, 15). Agriculture industry officials were also well represented in the trade delegations sent to negotiate the AoA (Desmarais 2003).

Liberalizing agricultural trade through GATT negotiations was also supported by important think tanks throughout the 1970s. For example, in 1973 the Brookings Institution produced a report that called for liberalizing agricultural trade during the Tokyo Round GATT negotiations (Brookings Institution 1973); and in 1974 the Trilateral Commission, whose members filled 25 prominent positions in the early Carter administration (Sklar 1980), also released a report highlighting the importance of including agriculture in the Tokyo Round negotiations (Paliano et al. 1977). Finally, in 1977 the American Enterprise Institute held a wide ranging conference on food and agricultural policy that included a session on world food markets and agriculture (AEI 1977). Session speakers and discussants all supported liberalizing agricultural trade through GATT negotiations, including Paul H. Trezise, senior fellow at the Brookings Institution and coauthor of the 1974 Trilateral Commission report, and Clayton Yeutter, assistant secretary of agriculture in the Nixon administration, deputy special representative for trade negotiations under President Ford, and future head of the U.S. Trade Representatives Office, where he led the U.S. negotiating team during the Uruguay Round of GATT negotiations from 1986 to 1989.

The evidence presented in this section thus demonstrates that by the early 1970s an important segment of the power elite had identified liberalized agricultural trade as an important U.S. policy objective, that agribusiness executives and members of the power elite conveyed this policy objective and strategies for achieving this objective to the Nixon administration through executive branch advisory committees that they sat on and directed (the Williams Commission and Flanigan's Council on International Economic Policy), and that the Nixon administration and every subsequent presidential administration has worked to liberalize agricultural trade through GATT negotiations. The evidence further demonstrates that agribusiness leaders played key roles in the GATT negotiations that produced the AoA and that in the 1970s several important policy planning organizations supported liberalizing agricultural trade through GATT negotiations, including the Trilateral Commission, which had very strong ties to the Carter administration (Sklar 1980), and the American Enterprise Institute, which had similarly strong ties to the Reagan administration (Peschek 1987). It is thus clear that the social and environmental consequences of liberalized agricultural trade—greater corporate control over agriculture, increased poverty for farmers, the erosion of national food self-sufficiency, the loss of millions of small and medium-sized farms, the increased mechanization of farming, reduced seed and crop diversity, and the increased use of petroleum and petrochemicals, etc.—are a direct outcome, at least in part, of economic policies devised by members of the power elite and the agribusiness industry to improve their economic position at the expense (either intentionally or unintentionally) of farmers and consumers around the world.

Why Developing Nations Accepted the Agreement on Agriculture

Given the highly detrimental effect the AoA has had on developing nation farmers, an obvious question that must be asked is, why did developing nations agree to the terms outlined in it? Many factors shaped their decisions to do so, but several key ones stand out. First, developing nations generally lack the resources to effectively determine their best interests in GATT negotiations or to attend all the GATT negotiating sessions affecting them. For instance, while the U.S. and European Union each send hundreds of delegates to these negotiations, developing nations are often able to send only a handful of representatives, and those representatives they do send often have limited resources with which to assess the different negotiating positions set forth at the meetings (Kwa 2002; Narlikar 2001). Second, during the Uruguay Round, developing nations were told that because of the competitive advantages they had in producing certain key agricultural products, they would benefit greatly from the liberalization of agricultural trade (Murphy 2009). Indeed, they were often told that while other aspects of the Uruguay Round agreements might not benefit them, the provisions set forth in the AoA would more than make up for these shortcomings (Jawara and Kwa 2003). Third, many developing nations are heavily dependent on the U.S. and other wealthy nations for export markets and various forms of aid. Combined with the demonstrated willingness of the U.S. to withhold resources from and otherwise punish developing countries that take different positions than the U.S. at WTO negotiations, this dependence often leads developing countries to go along with what the U.S. and other powerful nations want at these negotiations (Kwa 2002). Fourth, although decisions at the WTO and at WTO negotiations are supposed to be made by consensus, the consensus making process is set up so as to make it difficult for developing nations to build their own consensus or to oppose the decisions that wealthy nations want them to make (Kwa 2002; Narlikar 2001).

One of the main problems with consensus decision making for developing nations is that for consensus decision making to work for them, they have to publicly oppose the U.S. or Europe, the two most powerful actors at the WTO, if they disagree with either of them (the European Union—previously the European Community—negotiates as a single entity at GATT negotiations). However, if developing nations are unwilling to publicly declare their opposition to the U.S. or Europe, due to either fears of retaliation or a lack of resources with which to fully evaluate proposed treaty text, their lack of public opposition is treated as support for the proposed text. Developing nations also have trouble building their own consensus at the WTO because the U.S. and Europe are rarely in a position where they have to accept major treaty provisions they do not like. In addition, when developing nations do want to change the text of a draft proposal or declare their opposition to certain elements of it, they are often ignored by the officials running the negotiating session (Kwa 2002; Narlikar 2001).

Finally, the U.S. and Europe often negotiate key treaty text on their own, both prior to and during GATT/WTO negotiations, only bringing other nations into the negotiations after the U.S. and Europe have made most or all of the key decisions. Moreover, once they bring other nations into the negotiations, they often start by bringing in a handful of other wealthy nations and then a handful of more powerful developing nations, slowly building support from key nations so that by the time the negotiating text is made available to all the participating countries, those countries that were left out of the prior negotiations feel that they have been presented with a fait accompli that they have to accept. At this point, they also have fewer allies and less powerful allies to work with if they do want to question, criticize, or oppose the text they have been given (Kwa 2002; Narlikar 2001).

For all these reasons, as well as the fear of being left completely out of the international trading system, developing nations often agree to WTO treaty terms, including those set forth in the AoA, that are not in their best interests. Thus, GATT negotiating rules and the constraints that developing nations experience during GATT negotiations are key mechanisms, along with U.S. policy planning networks, power elite ties to the White House, and free trade agreements such as NAFTA and the WTO, that economic elites and powerful nations use to create and maintain an international trading system that benefits them at the expense of developing nations and the natural world.

International Patent Protection

International agreements that enshrine the U.S. interpretation of property rights and patent and copyright protection at the global level are another important mechanism used by powerful nations to increase the wealth and commodity chain power of economic elites and corporations. While these rights and protections have yet to be fully established globally, a key milestone in the move toward a U.S.-style global property rights regime occurred in 1994 with the ratification of the WTO's Agreement on Trade-Related Aspects of International Property Rights (TRIPs).

> TRIPS ... sets minimum standards in copyright, trade marks, geographical indications, industrial designs and layout-designs of integrated circuits. TRIPS [also] effectively globalizes the set of intellectual property principles it contains, because most states of the world are members of, or are seeking membership of, the WTO. It also has a crucial harmonizing impact on intellectual property regulation because it sets, in some cases,

quite detailed standards of intellectual property law. Every member, for example, has to have a copyright law that protects computer programs as a literary work as well as a patent law that does not exclude micro-organisms and microbiological processes from patentability. The standards in TRIPS . . . [thus] profoundly affect the ownership of the 21st century's two great technologies—digital technology and biotechnology. Copyright, patents and protection for layout-designs are all used to protect digital technology, whereas patents and trade secrets are the principal means by which biotechnological knowledge is being enclosed. TRIPS also obliges states to provide effective enforcement procedures against the infringement of intellectual property rights. (Drahos and Braithwaite 2002, 10–11)

Because the U.S., Europe, and Japan are the world's leading exporters of intellectual property, they and the corporations they represent have gained and will continue to gain the most financial benefit from TRIPs, in large part because TRIPs helps them maintain and exploit their intellectual property advantage (Chang 2008; Drahos and Braithwaite 2002, 2004). Indeed, one of the most important beneficiaries of TRIPs is the seed industry, which requires global property rights protections to profitably sell genetically modified seed around the world. It is unlikely, for instance, that without such protections the global area devoted to genetically modified crops would have expanded from zero to 395.4 million acres between 1996, the year genetically engineered seeds were first introduced commercially, and 2011, the most current year for which data were available when I wrote this, or that 19 of the 29 nations growing genetically engineered crops in 2011 would have been developing nations. Nor is it likely that without TRIPs developing nations would have planted nearly 50% of the global acreage devoted to these crops, that about 90% of the 16.7 million farmers growing these crops would have been poor, developing nation farmers, or that in 2010 "81% of all soybeans, 64% of cotton, 29% of corn and 23% of canola" grown globally would have been genetically engineered (C. James 2011; quote from Weise 2011).

Given the rapid expansion of acreage devoted to genetically modified crops—8% global acreage growth from 2010 to 2011, 10% growth the previous year, and 7% the year before that—and the fact that by 2010

10% of the world's farmland was used to raise these crops (C. James 2011; Weise 2011), it is likely that genetically engineered seeds will produce the same types of social and environmental problems globally as they have in the U.S. TRIPs is thus likely to produce, at the global level, higher seed prices for farmers, reduced farmer access to seed, a reduction in crop genetic diversity, increased herbicide use, herbicide resistance among weeds, reduced crop yields, and the transfer of genetically engineered material to non-genetically engineered crops, weeds, and wild species.[21]

Given the severity of these problems, the remainder of this section is devoted to briefly highlighting the mechanisms that corporate and government elites used to bring the TRIPs agreement into being. The story begins in the early 1980s when the pharmaceutical giant Pfizer realized that developing nations were beginning to use their superior numbers at the World Intellectual Property Organization (WIPO) to develop property rights rules that benefited them at the expense of large corporations. The solution that Pfizer adopted for this problem was to try to link intellectual property rights to international trade policy by including property rights as part of future GATT negotiations. To achieve this goal, the CEO of Pfizer, Edmund Pratt, began to garner support among other corporate executives by giving speeches at national and international trade association meetings and at meetings held by policy planning network organizations such as the National Foreign Trade Council and the Business Roundtable (other Pfizer executives did this too). Pratt also became a member and in 1981 chairman of the U.S. Advisory Committee on Trade Negotiations (ACTN), the most important of a set of committees set up in the 1970s to provide the U.S. Trade Representative's Office (USTR) with direct advice from U.S. industry regarding U.S. trade policy and international trade negotiations (Drahos 2003; Sell 2003).

Working closely with the two men who headed the USTR in the 1980s (William Brock and Clayton Yeutter), the ACTN was able to get the USTR to push hard for international property rights protections. Nevertheless, it became clear by 1986 that the U.S. would not get discussion of intellectual property rights onto the Uruguay Round negotiating agenda without support from Europe and Japan, which was not readily forthcoming. To change the position of European and Japanese negotiators, Edmund Pratt and John Opel, the chairman of the board and former president and CEO of IBM, formed the Intellectual Property Committee (IPC) in March 1986, "an ad hoc coalition of thirteen major US corporations [including] Bristol-Myers, *DuPont*, FMC Corporation, General Electric, General Motors, Hewlett-Packard, IBM, Johnson & Johnson, Merck, *Monsanto*, Pfizer, Rockwell International and Warner Communications" (Drahos 2003, 5; emphasis added). The CEOs of these corporations then successfully lobbied their European and Japanese counterparts, who, in turn, successfully pressured their governments to include intellectual property rights in the Uruguay Round negotiations (Drahos 2003; Sell 2003).

Once the Uruguay Round negotiations commenced in 1986, the kinds of tactics used to secure approval of the AoA were also used to secure approval of TRIPs. In particular, the U.S., Europe, and Japan made the most important negotiating decisions among themselves (May 2007), bringing other nations into and leaving them out of the negotiations as required and slowly bringing more nations on board as their opposition to the agreement was overcome (Drahos 2003; Drahos and Braithwaite 2002). The U.S. also placed extreme pressure on developing nations that opposed the TRIPs agreement, particularly on those that voiced the strongest opposition to the agreement, using the USTR's power to impose trade sanctions on developing nations to ultimately force these nations to accept TRIPs (Drahos 2002; May 2007). Finally, the Intellectual Property Committee provided significant legal counsel to the U.S. negotiating team throughout the TRIPs negotiations (May 2007) and, along with corporate leaders from Japan and Europe, wrote a draft property rights agreement (Drahos 2003) that with some "fine-tuning and concessions to developing nations" became the TRIPs agreement (May 2007, 29).

Conclusion

This chapter highlighted eight broad mechanisms that agribusiness firms, food retailers, and the U.S. government use to achieve the food and agriculture industry's capital accumulation goals: oligopoly and oligopsony power in agricultural commodity chains; liberalized agricultural trade; expansive property rights protections; free trade agreements that promote corporate-friendly property rights and liberalized agricultural trade; the organizational links that exist between members of the power elite that allow them to mobilize corporate and power elite support for specific policies;[22] the organizational links that exist between members of the power elite and the U.S. government that allow the power elite to strongly shape U.S. policy; decision making rules at the WTO that provide the U.S. and Europe with overwhelming advantages during GATT/WTO negotiations; and the use of economic threats by the U.S. to ensure cooperation from developing nations during GATT/WTO negotiations.

This chapter also highlights a set of commodity chain specific mechanisms that oligopoly and oligopsony firms in the agriculture industry use to undercut market competition, shape the behavior of other market actors, distort markets and prices to their benefit, and force buyers and sellers to accept terms of exchange dictated by the oligopoly and oligopsony firm. These commodity chain specific mechanisms include production and marketing contracts, commodity futures markets, the division of agricultural market exchanges across multiple cash and contract markets, informational advantages that oligopoly and oligopsony firms possess in agricultural markets, and the lack of viable cash markets for many agricultural goods. They also include seed trait licensing agreements, the ability of large seed companies to restrict the supply of conventional seed, technology agreements that farmers who purchase genetically modified seed must sign, investigations of and lawsuits against farmers who may or may not have violated these technology agreements, supermarket-led captive value chains, the heavy subsidization of crops used to produce nutritionally deficient food, and the ability of oligopoly firms to flood markets and sell goods at below cost.

Finally, this chapter demonstrates that the organizational, institutional, and network-based mechanisms highlighted in the chapter work together to increase profits for agribusiness firms and food retailers, severely harm farmers in developed and developing nations, devastate the environment, increase global hunger, reduce national food self-sufficiency, and harm consumers, in large part by providing agribusiness firms and food retailers with the power to strongly shape and in many cases dictate the crop choices, farming practices, and technologies that famers use and thus the food choices available to hundreds of millions, perhaps billions, of consumers.

In demonstrating how the mechanisms highlighted in this chapter work and how they harm people, communities, and the environment, I have provided strong support for my argument that environmental degradation and social harm are to a significant degree the product of undemocratic organizational, institutional, and network-based mechanisms that elites use to achieve specific capital accumulation goals, and thus the product, more generally, of organizational, institutional, and network-based (OINB) inequality.[23] I have also provided strong support for three of my six IDE hypotheses, in particular, for my predictions that elite-controlled organizations, institutions, and networks are a fundamental source of social and environmental degradation because they provide elites with the means to (a) monopolize decision making power, (b) shift environmental and non-environmental costs onto others, and (c) restrict the ability of non-elites (in this case farmers and consumers) to behave in environmentally sustainable ways by limiting the choices and shaping the incentives available to them.

Moreover, in providing broad empirical support for these hypotheses, I have demonstrated not only that the elite-controlled mechanisms examined in this chapter play a theoretically similar role in producing social and environmental harm as do the elite-controlled mechanisms described in the preceding and following chapters, but also that elites rely on multiple interacting and intersecting mechanisms, operating both within and across nations, to achieve their socially and environmentally harmful capital accumulation goals. This suggests quite strongly that activists and environmentalists will fail to solve the world's myriad social and environmental crises if they restrict their efforts solely to incremental policy fixes, to changing consumer behavior, or to incorporating social and environmental costs into the prices of the goods and services that businesses and consumers purchase. Indeed, it seems likely that radical action directed toward drastically reducing OINB inequality and dramatically reconstructing or abolishing elite-controlled organizations, institutions, and networks is necessary if environmentalists and activists are to successfully achieve their goals.

5

Armed Violence, Natural Resources, and the Environment

In chapters 3 and 4 I examined a variety of organizational, institutional, and network-based mechanisms, including the World Bank, structural adjustment, agricultural commodity chains, and liberalized agricultural trade, and demonstrated that these undemocratic and elite-controlled mechanisms play a key role in promoting capital accumulation, environmental degradation, and human suffering.[1] I shift my attention in this chapter to the role that armed violence plays in producing these same outcomes. Specifically, I demonstrate that armed violence is one of several overlapping and mutually reinforcing elite-controlled mechanisms that provide core nations and corporations with the means to control or gain disproportionate access to the natural resource wealth of developing nations, thereby promoting capital accumulation and military power in the core, degrading local, regional, and global environments, promoting human suffering, and creating the conditions within which the unequal international exchange of natural resources (ecological unequal exchange) can occur.

I set forth two broad arguments in this chapter. First, I argue that the organizations, institutions, and networks within which elite-controlled mechanisms are housed tend to be violent in at least one of three ways. They either (a) use or threaten to use armed violence to achieve their goals, (b) achieve their goals by relying at least in part on armed violence carried out by other organizations, institutions, or networks, or (c) severely harm, either physically, emotionally, or psychologically, individuals, communities, societies, and/or the natural world. I have, of course, already demonstrated that the elite-controlled mechanisms highlighted in the preceding chapters fall into the third category of violent mechanism: they directly, indirectly, or in conjunction with other elite-controlled mechanisms produce severe poverty, hunger, disease, disorder, dislocation, fear, trauma, etc. in communities and societies around the world.[2] This chapter therefore investigates elite-controlled mechanisms that fall into the first or second category of violent mechanism.

My second broad argument is that to fully explain the global ecological crisis, researchers need to understand, at a minimum, why wealthy industrial societies are able to extract so many natural resources, at such great volume, from so many places around the world. Obviously, extraction, transportation, communications, and information technology play an important role in allowing this to occur. However, the ability of wealthy industrial societies to extract vast quantities of natural resources from the earth and to get these resources to the factories, farms, and people that need them also requires complex social, economic, and legal arrangements that are instituted, organized, shaped, directed, and controlled by specific organizations, institutions, and networks (Bunker 2005; Bunker and Ciccantell 2005).

One set of institutions that facilitates resource extraction activities are international trade and finance institutions such as the World Bank, IMF, and WTO. These highly coercive institutions harm individuals, societies, and the environment in ways that are often violent in their consequences (Bello et al. 1999; Wallach and Woodall 2004; also see chapter 3), thereby placing these institutions and the elite-controlled mechanisms that shape their policies in the third category of violent mechanism I just discussed. However, these institutions do not directly use or control the means of armed violence, which I argue play a critical role, along with many of the elite-controlled mechanisms I have already discussed, in maintaining and increasing global resource extraction and ensuring the safe transport of raw materials and finished products. The means of armed violence are, instead, controlled by military, police, mercenary, and rebel forces around the world that are usually, but not always, associated with local or national governments and that sometimes act on their own behalf and sometimes to ensure capital accumulation. Thus, any complete explanation of the global ecological crisis and its relationship to human suffering and social degradation must focus theoretical and empirical attention on three issues: the relationship between resource extraction, raw material transport, armed violence, and environmental degradation; the structural role that military, police, mercenary, and rebel forces play in harming the environment; and the relationship that exists between elite-controlled organizations, institutions, and networks that directly control the means of armed violence and those that do not.

The chapter is organized as follows. I begin with a brief summary of sociological research on the relationship between armed violence, capital accumulation, and environmental degradation and identify several gaps in this literature that the chapter attempts to fill. I then explain why natural resource extraction plays a critical role in promoting global environmental degradation and why armed violence is likely to play an equally important role in ensuring that the secure extraction and transport of natural resources occurs. Finally, I demonstrate that armed violence is, in fact, strongly associated with natural resource extraction in many parts of the world.

To demonstrate this association, I identify 10 minerals that are critical to the functioning of the U.S. economy and/or military and then ask whether the extraction of these critical minerals involved the use of armed violence at any point in the 10–15 years before I began conducting this research. I define armed violence as violence and threatened violence perpetrated by military, police, mercenary, and rebel forces, and thus investigate violent acts such as military and police forces beating, arresting, or firing weapons at protestors, the use of mercenaries to provide mine security, the forced removal of local populations, and the use of forced labor to carry out resource extraction activities. I supplement this descriptive, but decontextualized, analysis with a set of short case studies that more fully examine the violent activities associated with the extraction of two of the 10 critical minerals included in the analysis (manganese and copper) and then briefly discuss examples of armed violence associated with the world's three largest mining companies and with petroleum and rainforest timber extraction. Finally, to demonstrate that armed violence works in conjunction with other elite-controlled mechanisms, I discuss several mechanisms that the World Bank uses to promote natural resource extraction in Africa and investigate the degree to which armed violence carried out by military, police, mercenary, and rebel forces is associated with African mining projects that have received World Bank funding or World Bank investment guarantees.

In doing these things, I achieve three key goals. I address several important gaps in the environmental sociology literature, demonstrate that violence is central to the processes and mechanisms highlighted in this book, and provide strong support for my theoretical model (the IDE model) by (a) empirically confirming five of my six IDE hypotheses and (b) demonstrating that elite-controlled organizations, institutions, and networks that directly control the means of armed violence work in conjunction with those that do not to harm people, societies, and the environment.

I turn now to a brief discussion of the environmental sociology literature.

Armed Violence, Ecological Unequal Exchange, and Environmental Sociology

In recent years a new variant of world systems theory, ecological unequal exchange theory, has been developed to explain the unequal exchange of natural resources and environmentally degrading activities between nations that are differentially located within the international political economy (Bunker 1984; Emmanuel 1972; J. Rice 2009). This new theory holds that due to their position in the world system hierarchy, core nations (wealthy nations) are able to take advantage not only of the labor power but also of the natural resource wealth of periphery nations (poor nations) while simultaneously exporting many environmentally degrading activities (including mining and timber harvesting) to the periphery and preventing periphery nations from taking full advantage of their own natural resources and labor power (see chapter 2 for a brief description of world systems theory). This unequal exchange of pollution, labor power, and natural resources results, they argue, in underdevelopment, impoverishment, and environmental degradation in the periphery and increased wealth and economic power in the core (Jorgenson and Clark 2009).

Ecological unequal exchange theory has received much empirical support, with researchers convincingly demonstrating that structural position in the world system hierarchy does, in fact, play an important role in shaping nations' ecological exchange relations and, thus, the ability of nations to minimize environmental degradation within their borders and take economic advantage of their own and global natural resources (J. Rice 2009). Ecological unequal exchange researchers have also begun to specify the organizational and institutional mechanisms that make ecological unequal exchange possible. For example, scholars have identified agricultural research institutes (Foster 1994) and military coercion and war (Clark and Foster 2009; Jorgenson and Clark 2009; and York 2008) as important mechanisms promoting the transfer of natural resources from the periphery to the core. There is also widespread recognition that structural adjustment and external debt force many developing nations to export their natural resource wealth to developed nations (Bello et al. 1999; Clark and Foster 2009).

Nevertheless, relatively few ecological unequal exchange researchers have thought systematically about the organizational and institutional mechanisms that underlie ecological unequal exchange; and they and other environmental sociologists have paid only limited attention to the role that armed violence and militarism play in degrading the environment and securing core nations' access to developing nations' natural resources (Clark and Jorgenson 2012; Hooks and Smith 2005).

This inattention to armed violence, militarism, and the environment has begun to change in recent years. For instance, a handful of environmental sociologists now argue that military activity is a key driver of environmental degradation (Clark and Jorgenson 2012), that routine military activity, weapons production, and war consume vast quantities of natural resources and produce severe environmental problems that cannot be attributed solely to the requirements of capital accumulation (Hooks and Smith 2004, 2005; Jorgenson and Clark 2009), and that powerful nations use military force and coercion against militarily weak nations to maintain their disproportionate access to these nations' natural resources (Clark and Foster 2009; Foster 2008; Jorgenson et al. 2010; O'Connor 1998; York 2008), thereby enhancing ecological unequal exchange (Jorgenson and Clark 2009).

In setting forth these arguments, these scholars make several important contributions to the environmental sociology and ecological unequal exchange literatures. Nevertheless, in focusing their attention on military inequality between nations, military violence between nations, and the environmental consequences of weapons production, military activity, and war, these scholars ignore several important ways in which armed violence and armed conflict contribute to local, regional, and global environmental degradation. For instance, while I agree with the argument that the environmental consequences of war and militarism cannot be attributed solely to the requirements of capital accumulation, I also believe that it is a mistake to argue that war and militarism are caused primarily by the independent interests of the state (that is, the interests of the state independent from the interests of capitalists), which is, in fact, what

Gregory Hooks and Chad Smith (2005), the two main proponents of this argument, seem to claim. Instead, it seems more reasonable to argue that in a capitalist world system violence and armed conflict ensure that natural resources vital to capital accumulation, core nation military power, and the state are extracted and transported in sufficient quantities and at low enough prices to (a) maintain or increase corporate profits and industrial production levels across core nation economic sectors, (b) guarantee levels of economic activity sufficient to maintain core nation tax revenues, and (c) provide core nations with the natural resources and tax revenues they need to maintain large, powerful militaries that, in turn, protect the interests of core nation capitalists and the state.

I further contend that the role armed violence and militarism play in ensuring core nations' access to natural resources is *not* restricted solely to war (or threatened war) between nations, though this is clearly a very important aspect of armed violence and militarism.[3] Among other things, armed violence also includes armed conflict between state and nonstate actors, violent repression carried out by state and rebel forces, the use of mercenary forces to protect natural resource assets, and the use and threatened use of military and police violence against protestors and local populations by states, corporations, and rebel groups.[4]

Thus, in the following sections of the chapter, I develop a theoretical argument that explains why armed violence, in its various manifestations, is likely to be one of several overlapping mechanisms that provide elite actors with the means to (a) prevail over others in natural resource conflicts, which include conflicts over the use, extraction, *and* transport of natural resources, and (b) ensure that natural resources critical to industrial production and state power continue to be extracted and sold in sufficient quantities to promote capital accumulation and state power in the core. I begin by explaining why developing such an argument is theoretically and substantively important. I then present the argument.

The Social and Environmental Significance of Natural Resource Extraction

The social, political, and economic importance of efficiently extracting and safely transporting natural resources cannot be underestimated. State, military, and geopolitical power, capital accumulation, social stability, industrial production, and the legitimacy of the state and economy all depend on large, increasing, and ever more concentrated withdrawals of natural resources from the earth (Bunker and Ciccantell 2005; Gould et al. 2004, 2008; Klare 2001, 2004; O'Connor 1996; Schnaiberg and Gould 2000).

However, these withdrawals also produce severe environmental degradation. Rainforest logging, for example, not only directly harms tropical rainforests, it also opens up vast areas of remote rainforest for ranching and agriculture, thereby directly and indirectly contributing to the destruction of a sizeable portion of the estimated 12.3 million hectares of tropical rainforest that are cut down each year (Cunningham and Cunningham 2007). Oil extraction also causes severe environmental problems, including deforestation, biodiversity loss, the degradation of aquatic and marine ecosystems, decreased soil fertility, air pollution, the release of heavy metals, acids, and carcinogens into local waterways (billions of gallons of such waste were released in Ecuador alone between 1964 and 1990), and the unintentional release of between 750 million and 1.8 billion gallons of crude oil into the environment each year (Dabbs 1996; Epstein and Selber 2002).

Mines are also environmentally destructive. In the U.S., for example, the 81 mines included in the Environmental Protection Agency's Toxics Release Inventory (TRI) in 2009 disposed of or released into the environment 1.13 billion pounds of toxic material,[5] more than any other industry in the U.S. and roughly one-third of the total contributed by the other 20,716 industrial facilities included in the 2009 TRI (EPA 2010). Mines produce so much waste because mined ore contains only the barest traces of the minerals that industry needs. As a result, worldwide each ton of extracted copper produces an average of 110 tons of waste ore plus 200 tons of waste overburden (soil and surface rock), while each ounce of gold that is mined produces an average of 79 tons of total waste (Farrell et al. 2004), with Newmont's Batu Hijau mine in Indonesia producing 250 tons of rock and ore waste for each ounce of gold produced (Larmer 2009). This waste is highly toxic, as are the chemicals that are used to separate minerals from ores.

This toxic waste, which can include arsenic, selenium, lead, cyanide, and mercury, often ends up in the local environment, where it can remain harmful for thousands of years. One reason it ends up in the local environment is that the tailings ponds that hold much of this waste often leak or fail. In 1995, for example, the tailings dam at the Omai gold mine in Guyana failed, releasing 3.4 million cubic meters of cyanide-laced waste into the Omai River (Fields 2001). Another reason it ends up in the local environment is that mining companies often dump their waste directly into rivers and oceans (Farrell et al. 2004), as occurs at the Ok Tedi mine in Papua New Guinea, where roughly 200,000 tons of waste ore and rock are dumped into the Ok Tedi river every day (Farrell et al. 2004, 7).

Of course, mining and other natural resource extraction activities do not harm the environment only at the "point of extraction." Because natural resources are the ultimate source of all the energy and goods we produce, consume, and throw away, natural resource extraction harms the environment globally as well. This is an incredibly important point because it implies that the grave environmental problems associated with industrial production and consumption (in both capitalist and noncapitalist societies) would not exist, or would not exist in their current form or at their current levels, if industrial societies were unable to efficiently extract and safely transport vast quantities of natural resources.

Computer production, for example, could not occur without the extraction of minerals, fossil fuels, and other natural resources from around the world. One such category of minerals is rare earth minerals, which are mined primarily in China (NRC 2008). The mining of rare earth minerals produces as much as 2,000 tons of solid waste, including toxic heavy metals and radioactive thorium, for every ton of rare earth mineral produced (Farago 2009; Rong and Yu 2009). In China it also results in topsoil loss, erosion, and the widespread silting and contamination of rivers and reservoirs used for drinking and irrigation (Xu and Liu 1999).

Tantalum is another mineral critical to computer production (NRC 2008). Tragically, the global market for tantalum helped fuel a six year war in the Democratic Republic of Congo that resulted in millions of deaths (United Nations 2002) and extensive deforestation, soil erosion, water pollution, and wildlife loss (Harden 2001).

In addition to harming the environment at the point of extraction, computers also harm the environment during the production, assembly, consumer use, shipping, disposal, and recycling stages of their lives, thus affecting the environment and human health around the world.

Environmental impacts during these stages of a computer's life include abiotic depletion, global warming, the release of toxins into the environment, human exposure to highly toxic materials, acidification, ozone depletion, the formation of photo-oxidants, and water eutrophication (Choi et al. 2006). In addition, because computers are so critical to globalization and economic growth, they help foster environmental problems associated with these phenomena too.

It would be difficult to argue, therefore, that the environmental problems associated with computer use and production are confined solely to the resource extraction stage of the commodity chain or that the environmental problems associated with the remaining stages of a computer's life would exist without the extraction of the minerals, fuels, and other natural resources needed to produce, ship, use, recycle, and dispose of computers. This is true, of course, of all the products we use and produce, including weapons systems, automobiles, solar panels, and cell phones. Thus, resource extraction is a pivotal link in the chain connecting human activity and social organization to local *and* global environmental degradation.

Given the fundamental importance of natural resource extraction and transport to capital accumulation, state power, and environmental degradation, as well as the importance that control over natural resources plays in shaping local, national, regional, and global distributions of power (Bunker and Ciccantell 2005; Klare 2001), it is imperative that environmental activists and researchers develop theoretical arguments that explain how powerful actors simultaneously attempt to ensure both their control over natural resources and the efficient and affordable extraction and transport of these resources. In the following section I develop such an argument.

Before proceeding, however, it is important to note that in arguing that armed violence plays a critical role in ensuring capital accumulation and state power, I do not mean to imply that core and periphery nations use or support resource-related violence for the same reasons or that states, rebels, mercenaries, and corporations that support resource extraction activities all have the same motivations for doing so. I discuss the implications of this in some detail in a subsequent section of the chapter. The important point for now is that the validity of the theoretical argument I present in the following section does not depend on these different groups of actors following the same decision making logic or having the same set of motivations for supporting resource extraction.

The Theoretical Argument

In developing my theoretical argument, I start with the assertion that those groups that successfully control, extract, and transport natural resources in specific places at specific times are able to do so because they have created a set of overlapping institutional, organizational, ideological, legal, and technological mechanisms that provide them with the means to prevail over others in conflicts over natural resource use, extraction, and transport. Stephen Bunker and Paul Ciccantell (2005) demonstrate, for example, that over the past 600 years, breakthroughs in transportation technology have resulted in dramatically increased levels of natural resource extraction, resource extraction from ever more remote locations of the world, and trade dominance for those nations that control the new technology and, as a result, have the greatest access to inexpensive natural resources. They further argue that the development of new transportation technology is contingent on the creation of elite-controlled financial, organizational, and institutional mechanisms that channel resources and structure relations "within and across firms, sectors, and the state" in such a way as to mobilize the resources needed to create technological breakthroughs and fund large-scale infrastructure development (Bunker and Ciccantell 2005, 18).

Of course, transportation technology alone does not ensure access to affordable raw materials, especially when those materials are located in other nations or in areas controlled by other groups. Thus, since World War II and the breakup of Europe's colonial empires,[6] wealthy nations and corporations, which consume the bulk of the world's natural resources (Hawken et al. 1999; Speth 2005), have relied on a combination of mechanisms that they control, including ideology (e.g., neoliberalism; Goldman 2005), debt (Bello et al. 1999; Clark and Foster 2009), agricultural research institutes (Foster 1994), export credit (Evans et al. 2002), political risk insurance (Moody 2005, 2007), and the World Bank, IMF, and WTO (Bello et al. 1999; Goldman 2005; Wallach and Woodall 2004; also see chapters 3 and 4 of this book), to maintain their access to and control over vital raw materials.

Among other things, these institutions, organizations, and ideologies have worked collectively to open the economies of developing nations to corporate investment, to increase the flow of natural resources from developing to developed nations, to garner developing nations' support for resource extraction and other corporate activities within their borders, and to create new legal structures and government institutions in developing nations that facilitate resource extraction and foreign economic involvement in these nations (Bello et al. 1999; Goldman 2005; Harrison 2004; Moody 2007; SAPRIN 2004; Toussaint 2005; Wallach and Woodall 2004; also see chapters 3 and 4 of this book).

However, nations and societies are not monolithic entities, and regardless of whether a government willingly or unwillingly engages in specific resource extraction activities, whether these activities are organized by local or foreign companies, or whether they occur in developed or developing nations or in nations with strong or weak legal and property rights regimes, it is likely that in many cases individuals and groups will protest, resist, or rebel against these activities. For example, protestors might be worried about local environmental degradation or health problems that result from resource extraction activities, they might be aggrieved by any loss of livelihood that they and their community may experience as a result of these activities, or they may be forced to relocate in order to make way for resource extraction (SAPRIN 2004). Similarly, workers hired by resource extraction firms may protest poor working conditions, local residents may receive few of the benefits but all of the burdens associated with resource extraction activities, or local residents may be indigenous, colonized, or otherwise marginalized people who resent government and outsider intrusion into their lives (Evans et al. 2002; Gedicks 2001; Moody 2007).

In such instances local and national governments, resource extraction firms, or rebels who control natural resources may feel that they have no choice but to use violence or the threat of violence to protect their resource extraction activities. Violent actions and threats of violence might include the forced relocation of local residents; the use of police, military, or mercenary forces to break up protests, arrest protestors, and provide mine security; or the repression of local indigenous people from whose ranks protestors have emerged or might emerge. Violent actions might also include military conflict with groups that threaten resource extraction activities or the provision of foreign military aid and training to local police and military forces.

Of course, armed violence may occur even in the absence of protest. For instance, forced labor may be used to decrease labor costs or because working conditions are horrendous, and the forced removal of people from their homes may occur in the absence of protest either to forestall protest or because there is no way to extract resources with people living on or near the extraction site. In either case, violence or threatened violence will likely be necessary because most people do not want to be forced to work or to leave their homes.

The use or threatened use of violence to gain or maintain access to vital raw materials may also occur in situations in which a resource that is viewed as being critical to national survival and economic prosperity, such as oil or water, is located in an area controlled by others in which mechanisms such as trade liberalization and structural adjustment have not effectively guaranteed permanent supplies of or control over the resource (for detailed discussions of this type of situation, see Klare 2001, 2004). In such situations, governments that want access to or control over the resource might engage in actions such as militarily enforced trade sanctions, proxy wars, counterterrorism, military threats, invasion, or providing military aid and training to local police and military forces.

Finally, because land and water transport are potentially subject to piracy and military disruption, governments may devote military resources toward protecting specific resource shipments (such as U.S. protection of Kuwaiti shipping during the Iraq-Iran war in the 1980s) or providing security for roads, railroads, and naval shipping lanes (Klare 2001, 2004).[7]

Corporations, States, and Rebels

Central to my theoretical argument in this chapter is the position I take on the question of whether core states, periphery states, corporations, and rebels share the same interests regarding natural resource extraction, resource-related violence, and capital accumulation. I expect that in most instances rebels, corporations, and states support resource extraction activities and use violence or the threat of violence in association with these activities in order to achieve their own goals, such as enriching themselves, increasing their power over others, maintaining social stability, or securing good relations with other actors. It is true, as I have argued throughout the book, that states face structural constraints that shape their behavior (see chapter 2 for a definition of social structure) and that government policy is shaped in critically important ways by powerful economic actors (Bello et al. 1999; Burris 1992; Domhoff 1990; Dreiling 2001; Toussaint 2005; also see chapters 3, 4, and 6). Nevertheless, it is also true that state interests are not fully reducible to capitalist interests or to structural constraints imposed on them by others (Hooks and Smith 2004; M. Mann 1986). Thus, when states use armed violence to support resource extraction activities, it is likely that their interests and goals are not identical to those of capitalists even when their policies, goals, and interests have been strongly shaped by capitalists.

But regardless of whether states that support natural resource extraction and resource-related violence do so for their own reasons, because capitalists want them to do so, or because they face structural constraints imposed on them by others, their decision to use or support the use of armed violence is still implicated in environmental degradation and capital accumulation if the violence or threatened violence contributes to these outcomes. In other words, states and corporations do not need to follow the same decision making logic or be motivated by the same set of interests for state violence to support capital accumulation and environmental degradation. The same is true for resource-related violence perpetrated by rebels and mercenaries.

Developing nations, of course, have very different interests than do core nations. They also have much less power than do core nations. As a result, in developing nations, the decision to use violence or the threat of violence in order to protect resource extraction activities *is* likely to be strongly shaped by structural constraints imposed on them by wealthy governments, corporations, and international institutions.

For example, structural adjustment programs imposed by the World Bank and IMF often force developing nations to maintain high levels of raw material exports (see chapter 3); and in cases where mining projects require political risk insurance, developing nations are sometimes forced to agree that they will pay out potentially large insurance claims if mining activities are disrupted in any way (Moody 2005, 2007).[8] Developing nations' high levels of debt and their resulting dependence on wealthy nations, the World Bank, the IMF, and corporate foreign investment also force developing nation governments to worry about how these powerful organizations and states will evaluate their activities (for instance, see chapters 3 and 4). As a result, developing nation governments may feel that regardless of their own motives and interests, they have to use all means necessary to protect resource extraction activities so as to meet their debt obligations, ensure continued foreign investment, and minimize conflict with more powerful nations and institutions.

Moreover, even in cases where rebels and developing nation elites benefit from resource extraction activities, and thus find these activities to be in their interests, it is still the case that they are junior partners in the commodity chains that link resource extraction to global production and consumption. As a result, they have relatively little power in these commodity chains and their interests and behavior in these chains are strongly shaped by more powerful commodity chain actors as well as by the core states and international institutions that support these powerful actors.[9]

I therefore contend that the use of armed violence by developing nations to protect resource extraction activities is likely to be strongly conditioned by the overlapping institutional, organizational, ideological, legal, and network-based mechanisms that I previously argued provide corporations and wealthy nations with the means to control natural resources and prevail over others in conflicts over these resources. If this is correct, it implies that armed violence is one of these important overlapping mechanisms even when developing nations make the actual decision to use this violence. In other words, if corporations and wealthy nations play a key role in shaping the conditions within which developing nations and rebels decide to extract large quantities of natural resources, and if extracting and transporting these resources requires the use of armed violence by these nations or rebels, then core nations and corporations are relying on this violence, which is a "local" elite-controlled organizational mechanism, to achieve their global capital accumulation goals.

We can thus think of the overlapping mechanisms that ensure core nations' control over natural resources as operating at multiple geographic levels and as being nested within each other such that the mechanisms that are employed within developing nations, such as armed violence and corporate-friendly property rights regimes, are strongly conditioned by mechanisms such as international debt and World Bank–enforced structural adjustment programs that operate at the international level. Moreover, we can work backward from global institutions such as the World Bank to the various organizational mechanisms that the U.S. uses to shape the policies of international institutions and that members of the U.S. power elite use to shape U.S. policy vis-à-vis these institutions (see chapter 3), in this way tracing the power of the U.S. power elite through a set of undemocratic organizational mechanisms that originate within the U.S., move to the international arena, and then settle back down in specific developing nations.

Mining and Violence

In the preceding pages I argued that armed violence is one of several overlapping mechanisms that provide powerful actors with the means to prevail over others in conflicts over natural resources and ensure that natural resources that are critical to industrial production and state power continue to be extracted and sold in sufficient quantities to promote capital accumulation and state power in the core. If this argument is correct, it suggests that natural resource extraction and transport are likely to be associated with armed violence in many instances, but that the successful operation of other "resource extraction" mechanisms, such as property rights regimes, debt, and structural adjustment, will often preclude the need for armed violence.

In order to empirically evaluate the argument that natural resource extraction is often associated with armed violence, I now ask whether armed violence is or has been associated with the extraction of 10 minerals that are critical to the U.S. economy and/or military and, if so, what types of armed violence have been associated with the extraction of these critical minerals. However, before proceeding, it is important to note that I conducted the research for this investigation in 2009 and that the evidence presented in this chapter differs from the evidence presented in other chapters in that much of it is drawn from newspaper accounts of specific events. As a result, the evidence in this section is based on events that occurred prior to 2009, and I spend several pages discussing how my research assistant (Eric Bonds) and I collected and categorized the newspaper data and defined the terms that we used when conducting the study.

It is also important to note that I do not attempt to show that ecological unequal exchange occurred for the minerals examined in this chapter because other researchers have convincingly demonstrated that natural resources and natural resource wealth flow unequally from periphery to core nations (J. Rice 2009).[10] As a result, all I need to do to demonstrate that armed violence plays an important role in ensuring that ecological unequal exchange occurs is to show that armed violence—which I define as violence and threatened violence perpetrated by military, police, mercenary, and rebel forces—is strongly associated with the extraction of a wide variety of natural resources.

The 10 minerals I investigate in this section of the chapter are platinum, palladium, rhodium, manganese, indium, niobium, vanadium, titanium, copper, and rare earth elements (platinum, palladium, and rhodium are collectively referred to as platinum group metals). I selected these minerals for examination based on the findings of a National Research Council study (NRC 2008) that evaluated how critical each of a larger group of minerals is to the U.S. economy and U.S. national security. The NRC, in turn, based its evaluations on quantitative and qualitative measures that included for each mineral in its study the proportion of that mineral's U.S. consumption that is devoted to producing specific types of products (such as electrical components, aerospace materials, integrated circuits, and energy provisioning materials), the overall economic importance of "the industrial sector encompassing the dominant use of the mineral" (NRC 2008, 112), the importance of the mineral's end-use products to U.S. military preparedness, and how difficult and expensive it would be to provide an equivalently useful substitute for the mineral if its price increased dramatically or its supply was cut drastically.[11]

Because each mineral in the NRC report has multiple end uses, I selected for examination each of the minerals in the report that received the report's highest "supply disruption" score for at least one of its end uses. These scores range from 1 to 4, with a score of 1 indicating that a supply disruption would have little impact on the U.S. economy or national security and a score of 4 indicating a high level of potential disruption due to a lack of suitable alternatives for that particular end use.

Table 5.1 lists the 10 minerals, the end uses for which each mineral received a supply disruption score of 4, the nations that had the largest reserves and highest production levels of each mineral in 2006, and (in

TABLE 5.1. Critical Mineral End Uses, Production, and Reserves

Mineral	Use with highest impact score	Percentage of world production	Percentage of world reserves
Platinum	Autocatalysts for motor vehicle emissions control; industrial and other applications	South Africa (77%), Russia (13%), Canada (4%), Zimbabwe (2%), U.S. (2%)	Total platinum group metal reserves: South Africa (89%), Russia (8%), U.S. (1%)
Palladium	Autocatalysts for motor vehicle emissions control; industrial and other applications	Russia (44%), South Africa (38%), U.S. (6%), Canada (6%), Zimbabwe (3%)	(see platinum)
Rhodium	Autocatalysts for motor vehicle emissions control; industrial and other applications	South Africa (89%), Russia (7%), North America (2%)	(see platinum)
Rare earth elements	Emissions controls, magnets, and electronics	China (76%), France (9%), Japan (4%), Russia (3%)	China (31%), Commonwealth of Independent States (22%), U.S. (15%), Australia (6%), India (1%)
Manganese	Steel and cast iron	South Africa (19%), Australia (18%), China (13%), Brazil (12%), Gabon (11%)	Ukraine (30%), South Africa (22%), Australia (13%), India (12%), China (9%)
Indium	Coatings (flat-panel displays, etc.)	China (60%), Japan (9%), South Korea (9%), Canada (9%), Belgium (5%)	China (73%), Peru (3%), U.S. (3%)
Niobium	Superalloys (for the aerospace industry, etc.)	Brazil (90%), Canada (9%)	Brazil (96%), Canada (2%), Australia (2%)
Vanadium	Nonsteel alloys	South Africa (39%), Canada (31%), Russia (27%)	China (38%), Russia (38%), South Africa (23%)
Titanium	Aerospace and high technology	Australia (27%), South Africa (20%), Canada (14%), China (9%), Ukraine (6%)	China (27%), Australia (20%), India (13%), South Africa (10%), Brazil (6%)
Copper	Energy provisioning and power production	Chile (35%), US (8%), Peru (7%), China (6%), Australia (6%)	Chile (31%), US (7%), Indonesia (7%), Peru (6%), Mexico (6%), Poland (6%)

parentheses) the percentage of reserves and production that each of these nations contributed to the world total.[12] The table indicates that the 10 minerals play a critical role in the production of a variety of economically and militarily important products, including steel, cast iron, superalloys and nonsteel alloys (such as those used in the aerospace industry), coatings (including for flat panel screens), electronic devices, aircraft, autocatalysts for motor vehicle emissions control, magnets for digital storage and other high-tech applications, medical products, and power production.

The table also indicates that in 2006 six of the 10 minerals were produced primarily in Russia, China, and/or South Africa, with Brazil, Canada, Australia, and Chile producing large amounts of some of the minerals and other nations producing smaller amounts of some of the minerals. The geographic distribution of the minerals included in the table results in part from the NRC's concern with minerals that are located in unstable regions of the world or in nations that are more likely to withhold resources from the U.S. (Russia and China) and may bias, to some degree, the strength of the findings presented in table 5.2. I attempt to compensate for this potential bias by recording acts of armed violence that occur or have occurred in nations that are both major and minor suppliers of these minerals and by discussing, in subsequent sections of the chapter, violence related to the world's three largest mining companies and to the extraction of other critical minerals and raw materials.

In order to examine the link between mining and armed violence, table 5.2 indicates whether armed violence is associated with the extraction of any of the 10 minerals included in the study and, if so, what form this violence takes and in which country or countries the violence occurred (in table 5.2, the three platinum group metals are listed as a single entry). Table 5.2 lists seven forms of armed violence and threatened violence, which include the following: direct military or police actions have been taken against anti-mine protestors or rebels; military, police, or mercenary forces provide mine security; the government is highly repressive; mining operations have resulted in the repression of indigenous or colonized peoples; local residents have been forced to leave their homes to make way for the mine (forced removal); forced labor or prison labor has been used at the mine or to make the mine operational; and "other forms of armed violence," a residual category that includes forms of armed violence not captured by the other six categories.

TABLE 5.2. Critical Minerals and Violence

Mineral	Military, police, or mercenary actions against protestors or rebels	Military, police, or mercenary security of mines	Repressive regimes	Repress indigenous or colonized peoples	Forced removal	Forced/prison labor	Other violence
Platinum group	South Africa		Soviet Union/Russia (4.1) South Africa (3.3)		South Africa	Soviet Union	Russia (area closed to outsiders; journalists need permission to enter)
Rare earth elements	Malaysia		China (Inner Mongolia) (4.2) Russia (4.1) India (3.6)	China (Inner Mongolia)			Malaysia (1987: activist held for up to two months in solitary confinement, where he was interrogated and at one point kept awake for 78 hours)
Manganese	China Brazil (see other violence for description)		China (4.2) Brazil (3.9) Gabon (no data, see note 13) South Africa (3.3) India (3.6)		Brazil (as part of a railway project designed to gain access to the region's minerals)		Brazil (protesters were beaten, threatened, and sometimes tortured and killed) China (hired thugs beat protestors; in one town all men older than 16 were held for interrogation)
Indium	Tibet		China (Tibet) (4.2)	China (Tibet)			

183

TABLE 5.2. Critical Minerals and Violence *(cont.)*

Mineral	Military, police, or mercenary actions against protestors or rebels	Military, police, or mercenary security of mines	Repressive regimes	Repress indigenous or colonized peoples	Forced removal	Forced/ prison labor	Other violence
Niobium			Brazil (3.9)	Brazil	Brazil (as part of a larger state project to secure the border, fight drug trafficking, and facilitate resource extraction)		
Vanadium			China (4.2) Russia (4.1) South Africa (3.3)				
Titanium	Sierra Leone	Sierra Leone (mercenaries)	China (4.2) India (3.6) South Africa (3.3) Brazil (3.9)		Sierra Leone Kenya South Africa (potential)		Kenya (when lawyers tried to inform farmers of their rights regarding forced relocation, police told them to stop or they would be arrested) India (government was prepared to appropriate 10,000 acres of land for mining until protesters convinced the government to put the mining project on hold)

Mineral	Military, police, or mercenary actions against protestors or rebels	Military, police, or mercenary security of mines	Repressive regimes	Repress indigenous or colonized peoples	Forced removal	Forced/ prison labor	Other violence
Copper	Indonesia (West Papua) Papua New Guinea (war)	Indonesia (West Papua: The mine has paid the military and police for protection and provided the state with military infrastructure)	Indonesia (West Papua) (3.7) Papua New Guinea (3.5) Myanmar (4.1) Philippines (3.7) Mexico (3.2)	Indonesia (West Papua) Papua New Guinea (war)		Myanmar	Indonesia (West Papua kept off-limits to outsiders; military personnel have been accused of rapes, killings, and other abuses in defense of the mine; spying on environmentalists)

I include *military and police actions against protestors and rebels* and *repression of indigenous or colonized peoples* in the table because these are direct examples of armed violence being threatened or used to promote mining operations and because such actions imply the possible use of armed violence in the future, and thus also represent a threat of future violence. In addition, even after a group has been colonized, continued domination of the group is likely to be supported at least in part by armed violence or direct threats of armed violence. I include *forced labor, prison labor*, and *forced removal* in the table because these forms of coercion are all backed by the threat of armed violence and often by the actual use of armed violence, and I include *mine security by the military, police, and mercenaries* in the table because this use of military, police, and mercenary forces represents a clear threat of armed violence and sometimes involves the actual use of armed violence. The use of military and police personnel to provide mine security also indicates some sort of formal or informal agreement between the mining company and the state concerning the use of armed violence.

Finally, I include *highly repressive regimes* in the table because in such regimes all state supported activity (including resource extraction) is backed by the threat of armed violence. In addition, highly repressive regimes routinely use the threat of armed violence and imprisonment to forestall protest and minimize open political conflict, thereby reducing the likelihood that local residents will voice their grievances against mining operations. As a result, repressive regimes may not need to use direct acts of armed violence to maintain mining operations, relying instead on overall levels of violence and fear to achieve their mining goals. Because repressive regimes generally censor news accounts and restrict the movement of people and journalists within their borders, it is also the case that news organizations may have difficulty reporting accurately about these regimes' violent actions.

To determine whether governments are highly repressive, I use the Political Terror Scale (PTS), which is "generally considered to be the most valid and . . . widely employed [measure of repression] in the [political science] literature" (Davenport 2007, 79). The PTS rates countries on a scale from 1 to 5, with 5 being the most repressive, and I designate as highly repressive those countries with a score greater than or equal to 3.[13] A score of 3 indicates that "there is extensive political imprisonment, or a recent history of such imprisonment," that "execution or other political murders and brutality may be common," and that "unlimited detention, with or without a trial, for political views is accepted" (Gibney et al. 2009). A score of 5 indicates that "terror has expanded to the whole population" and that "the leaders of these societies place no limits on the means or thoroughness with which they pursue personal or ideological goals" (Gibney et al. 2009).

With the exception of the PTS scores, the information included in table 5.2 (and in the case studies presented in the following two sections) was obtained through archival research. Eric Bonds and I first consulted the U.S. Geological Survey's *2006 Minerals Yearbook* and *2008 Commodity Summary* (USGS 2006, 2008) to determine which countries were the major and minor producers of each mineral and to identify individual mines responsible for a substantial proportion of each mineral's world production. We then conducted a series of LexisNexis searches to determine whether armed violence is associated with the extraction of any of these minerals. We instructed LexisNexis to search in major world newspapers printed in or translated into English (the only language we know) and used search phrases that included terms such as "copper & Indonesia & violence," "manganese & Gabon & police," and "Grasberg & mine & military." In order to gather information unavailable in newspaper accounts, we supplemented the LexisNexis searches with Internet and library research as needed, utilizing websites such as those published by the following organizations: Mines and Communities (www.minesandcommunities.org), Business and Human Rights (www.business-humanrights.org), and Yale Environment 360 (www.e360.yale.edu).

It is important to note that we restricted the LexisNexis searches to acts of armed violence occurring in periphery and semi-periphery nations in the 10–15 years prior to 2009 (the year we conducted the research), though we sometimes learned of earlier acts of armed violence through our supplemental research. We restricted the searches to this time period in order to demonstrate that the use of armed violence has played a key role in recent years in ensuring core nations' access to critical natural resources and because LexisNexis rarely provided us with relevant articles further back than 10 years. We focused on periphery and semi-periphery nations since one of the main goals of this chapter is to contribute to the literature on ecological unequal exchange, which occurs between core nations on the one hand and periphery and semi-periphery nations on the other hand.[14]

It is also important to note that for several reasons the evidence we obtained using our archival search strategy likely underrepresents the extent to which armed violence is used to facilitate mineral extraction in periphery and semi-periphery nations. First, it is likely that acts of armed violence associated with mineral extraction often escape the attention of the world's largest newspapers. Second, because we restricted our search to major newspapers that publish in or are translated into English, we did not obtain evidence from smaller newspapers or from newspapers that are not published in or translated into English. Third, many of the world's major mineral producers are countries (such as China, Colombia, Indonesia, and Russia) in which the press enjoys limited freedom, making it less likely that violent actions taking place in these countries would show up in our search. Fourth, it was very difficult to obtain any information about two of the critical minerals we investigated (indium and vanadium) because they are both found in a broad range of mineral compounds, making it difficult to identify where they are mined.

Despite these limitations, table 5.2 still shows that armed violence is associated with the extraction of all the critical minerals examined in this chapter (though in the case of vanadium the violence solely takes the form of generalized state repression). For example, for these critical minerals, violent actions against protestors or rebels occurred in South Africa, Malaysia, China, Brazil, Tibet, Sierra Leone, Indonesia, and Papua New Guinea (these actions ranged from armed police facing off against protestors but not moving beyond the threat of violence to police arresting, beating, firing on, and killing protestors to open warfare against rebels). Mercenaries and military personnel provided mine security for these minerals in Sierra Leone and Indonesia, and the mining of these minerals occurred under repressive regimes in the Soviet Union, Russia, South Africa, Inner Mongolia, Tibet, China, Brazil, Gabon, Indonesia, Papua New Guinea, Myanmar, the Philippines, India, and Mexico, and involved the repression of indigenous or colonized people in Inner Mongolia, Tibet, Brazil, Papua New Guinea, and Indonesia. In addition, the extraction of these minerals involved the forced removal of people from their homes in South Africa, Brazil, Sierra Leone, and Kenya, the use of prison labor or forced labor in the Soviet Union and Myanmar, and the use of other forms of armed violence, including threatened arrests, solitary confinement, sleep deprivation, rapes, and killings, in Russia, Malaysia, Brazil, China, Kenya, and Indonesia.

These findings thus provide strong support for my argument that the extraction of critical natural resources is often, though by no means always, associated with armed violence (especially when we recall that it was nearly impossible to ascertain whether armed violence is associated with indium and vanadium mining). However, in presenting these findings in table form, I have only provided a partial understanding of the violent context within which the mining of critical minerals often occurs. Moreover, the evidence presented in table 5.2 does not allow me to evaluate any of my theoretical claims about why mining is likely to lead to violence.

Thus, in the following sections I more fully examine the violent activities associated with the extraction of two of the critical minerals highlighted in these tables: manganese and copper. I selected these critical minerals for closer examination because their extraction is associated with armed violence in multiple countries, because there are clear variations in the types and levels of armed violence associated with each of them, and because doing so allows me to examine several of my theoretical claims. Thus, my purpose in discussing them further is not to argue that mineral extraction is always associated with armed violence or with extremely high levels of armed violence but rather to more completely evaluate my theoretical argument and more clearly illustrate how integral armed violence can be to the extraction of critical mineral resources.

Manganese

Manganese is a widely used mineral essential for steel making. It is also used to produce nonsteel alloys and batteries (USGS 2006). According to the NRC (2008), the United States is almost entirely dependent on imported manganese, for which there are no known technical substitutes. In 2006 the leading producers of manganese ore were South Africa (19%), Australia (18%), China (13%), Brazil (12%), and Gabon (11%) (USGS 2006), with armed violence directly associated with manganese extraction in at least two of these countries (China and Brazil) and indirectly in a third (Gabon).

In Xiushan County, China, for example, there are 41 licensed and more than 200 unlicensed manganese mines, which in 2008 accounted for a large share of China's total manganese production (Jigang and Chuhua 2008). These mines have drained the region's aquifers and illegally dumped tons of toxic waste into the region's waterways, resulting in local wells running dry when rainfall is low, local rivers and irrigation water becoming severely polluted, and the region's rice harvest being cut in half. Constant mine blasts have also cracked the foundations of residents' homes and impaired local air quality (Jigang and Chuhua 2008).

In order to protect their livelihoods and health, some Xiushan County residents confronted local authorities in 2005 by blockading mine entrances and demanding the enforcement of environmental regulations. They also sought adequate compensation for their losses. One notable blockade, which lasted more than a month, was mounted by 40 elderly women (Jigang and Chuhua 2008). Mine authorities responded to these actions by hiring thugs to beat protesters, while Chinese police responded by raiding the town of Gaodong and detaining all townsmen older than 16 for interrogation (Jigang and Chuhua 2008).

Elsewhere in China, residents of Xialei, in Guangxi Province, conducted a nonviolent sit-in protest in 2006 to stop development of a local manganese electrolysis plant (Xi 2006). Because manganese mining already polluted the community's drinking water, the goal of the protest was to prevent further harmful pollution (Xi 2006). When two of the protesters were arrested, more than 1,000 Xialei residents gathered in front of the town hall to demand their release. In response, Chinese authorities mobilized several hundred police officers from surrounding communities, who beat the protestors and shocked them with electric batons (Xi 2006).

Manganese mining is also associated with armed violence in Brazil. Brazil's largest manganese deposit, in the Carajas region of the Amazonian Basin in Para state, is mined by Companhia Vale do Rio Doce (CVRD/Vale)[15] at its Azul mine (USGS 2006) and by Prometel Produtos Metalurgicos (PPM) at its Buritirama mine (*Mining Magazine* 1993). Prior to the 1970s the Carajas region was very remote and difficult to reach. However, the largest iron ore deposit in the world was discovered there in the 1960s, in 1971 manganese ore was discovered in the region, and in 1978 CVRD/Vale undertook a massive project, the Carajas Grande project, which received considerable state financing and foreign capital (including funding from the World Bank) to build a railway link between Carajas and the Atlantic Ocean (Shaw 1990).[16]

The Carajas Grande project, which was completed in 1985, opened up the Carajas region to iron, manganese, and copper mining as well as to other industrial activities and, as described in chapter 3, severely degraded the environment. It also led to the involuntary removal of many indigenous people from their land due to the granting of hundreds of prospecting concessions in the newly opened area (Rocha 1986) and a massive influx of miners, loggers, ranchers, and settlers. The massive influx of new people into the region not only pushed indigenous people from their land, it also made them the victims of violent attacks in the scramble for land and resources (Survival International 2000).

Peasants and indigenous people in the region also suffered violent attacks at the hands of the Brazilian government. For instance, upon completion of the Carajas Grande project, Brazil sold a great deal of land in the region to large corporations, including one parcel that it sold to CVRD/Vale. When CVRD/Vale evicted farmers from this parcel, the farmers mounted a road blockade to pressure the government and CVRD/Vale to provide compensation for their loss of land and livelihoods (Amnesty International 1998). Military police, who were transported to the area in CVRD/Vale vehicles, responded by beating and shooting the protesters, leaving 19 dead and wounding an unspecified number of others (Amnesty International 1998).

As noted in chapter 3, the success of the Carajas Grande project was also dependent on the construction of a large hydroelectric dam, the Tucurui dam, to provide power for mineral development (Hall 1989).[17] In order to make way for the dam, the Brazilian government expropriated the land of some 30,000 people, including land in two areas previously declared "indigenous reserves" (La Rovere and Mendes 2000). The dam degraded fisheries, flooded farmland, destroyed rainforests, created the conditions for malaria epidemics, and left thousands of people landless (La Rovere and Mendes 2000). A defoliant similar to Agent Orange was also used to clear a path for transmission lines to and from the dam, which some people believe increased the incidence of death, miscarriage, and illness in the area (La Rovere and Mendes 2000). Communities that protested against these conditions and demanded compensation for their losses were met with state repression, including one instance in which police attempted to disperse a large protest encampment with teargas and blocked some protesters' access to food, clean water, and other provisions (Hall 1989).

Thus, in both Brazil and China the government used armed violence to suppress activists who were protesting against the negative repercussions of manganese mining. Consistent with my argument, these negative repercussions included environmental degradation, health problems associated with mining and the use of defoliants, forced removal (in Brazil), the loss of livelihoods that resulted from forced removal, and the loss or degradation of fisheries, farmland, and rice harvests. Further supporting my argument, forced removal occurred in Brazil because large numbers of people lived in areas where mining concessions were granted or where flooding from the new dam was to occur.

The relationship between manganese mining and armed violence is less direct, but no less important, in Gabon, which produced 72% of the manganese used in the U.S. in 2006 (NRC 2008). For example, although I found no reports of armed violence specifically being used to support Gabon's manganese industry, Gabon is an undemocratic country with a ruling elite that controls the nation's military and paramilitary forces and benefits directly from the nation's mineral wealth. The Gabonese government, which had only one president from 1967 to 2009, also regularly violates its citizens' human and political rights. Violations include

the limited ability of citizens to change their government; the use of excessive force, including torture, on prisoners and detainees; harsh prison conditions; violent dispersal of demonstrations; arbitrary arrest[s] and detention; an inefficient judiciary susceptible to government influence; restrictions on the right of privacy; restrictions on freedom of the press, association, and movement; widespread harassment of refugees by security forces; widespread government corruption; violence and societal discrimination against women and noncitizen Africans; trafficking in persons, particularly children; and forced labor and child labor. (U.S. Department of State, 2006b)

Under such conditions, it is unlikely that citizens will be willing to protest local mining activities, giving the Gabonese government little reason to use armed violence to directly protect these activities. Nevertheless, because all state supported activity in highly repressive regimes such as Gabon's is essentially backed by the threat of violence, it is reasonable to argue that armed violence and the threat of armed violence play a critical role in facilitating Gabon's mining operations. Moreover, Gabon's restrictions on freedom of the press and freedom of movement mean that media coverage and human rights reports may inaccurately or incompletely portray the nation's mining-related activities. Amnesty International, for example, provided no information on human rights in Gabon at the time that we were conducting this research.

Copper

Copper is essential to the functioning of the U.S. and global economies due to its importance in the building and construction industries, the generation and transmission of electricity, and the manufacturing of transportation equipment, industrial equipment, and machinery. To some extent, substitute materials are available for some of these end uses, but there are no substitutes for electricity generation and transmission (NRC 2008), without which the U.S. and global economies would grind to a halt.

Nowhere is the association between copper mining and armed violence more apparent than in West Papua, Indonesia. Freeport-McMoRan, one of the largest copper and gold producers in the world, first began mining copper and gold in West Papua in 1972 but significantly ramped up its West Papuan production when it constructed its Grasberg mine in 1988 in partnership with the Indonesian government (Rio Tinto, which became a part owner of Freeport-McMoRan in 1995, sold its Freeport shares in 2004 but still retains a major joint venture interest in the mine). The Grasberg mine is one of the largest copper mines in the world, producing approximately 1,644 tons of copper and 700,000 tons of mine waste a day (Freeport-McMoRan 2007; Perlez 2006c).[18] This waste is carried by rivers to wetlands and estuaries, which at one time were among the most productive fisheries in the world. However, pollution from the mine waste caused massive fish die-offs, and few fish live in the polluted waterways today (Perlez and Bonner 2005).

Mining activities also resulted in the forced removal of indigenous people from their homes, the destruction of areas that hold religious and cultural significance to them, and economic hardship and loss of subsistence livelihoods for indigenous people who still live near the mine (Walton 2001).

West Papua has also been unwillingly subjected to Indonesian rule since the 1960s and is inhabited by individuals who have distinct native languages and different religious practices than do other Indonesians. Due to these and a variety of other factors, including widespread human rights abuses by the Indonesian army and the destruction of West Papua's environment and expropriation of its natural resource wealth for the benefit of others, West Papuans have long advocated independence from Indonesia, and some have undertaken a low-level insurgency to achieve this goal (Leith 2003; Perlez and Bonner 2005).

West Papuans have also protested and rioted in response to Freeport-McMoRan's mining activities, arguing that they have received little benefit from the mine while bearing most of its environmental costs (*New York Times* 2006; Perlez 2006a, 2006b). Anti-mine activists are also upset by the loss of land and subsistence livelihoods that have resulted from mining activities and by the dramatic growth in the area's population brought on by mining, which they argue has resulted in alcohol abuse, AIDS, prostitution, and fighting between soldiers and the police (Perlez and Bonner 2005; Walton 2001).

In response to this protest and rebellion, Freeport-McMoRan and the Indonesian government have worked closely with each other to suppress activists and rebels and protect the Grasberg mine and other mining activity. For instance, after insurgents sabotaged Freeport's copper slurry line in 1977, the Indonesian military "carpet bomb[ed], straf[ed], and reputedly napalm[ed] . . . surrounding villages" (Leith 2003, 226–227); and in support of the Grasberg mine, the Indonesian military has forcefully put down student riots, taken student anti-mine and pro-independence leaders into custody, conducted door-to-door raids, and at one point imposed military rule over the region surrounding the city of Jayapura, where thousands of student activists were protesting the devastating environmental impacts of the mine, its lack of local benefits, and the human rights abuses of the Indonesian military (Kearney 2006; Perlez 2006c).

The Indonesian military has also policed and protected the mine itself (Perlez 2006b) and has been accused of committing rapes, extrajudicial killings, and other human rights abuses in defense of the mine (Perlez and Bonner 2005). In return, Freeport-McMoRan has spent millions of dollars on the Indonesian military. For example, the company is known to have given at least $20 million in direct payments to the Indonesian military and police to protect the mine, though some people believe the actual figure is much higher, and as of 2005 the company had given the Indonesian military $35 million for military infrastructure (Perlez and Bonner 2005). Former company officials also claim that Freeport-McMoRan and the Indonesian military have worked together to spy on environmentalists concerned about the impacts of the mine (Perlez and Bonner 2005).

Copper mining is also associated with military activity and armed violence in other locations as well. For instance, Rio Tinto's experiences as a partner in the Grasberg mine are not so different from its experiences at its Panguna copper mine in the now autonomous region of Bougainville, Papua New Guinea. Although not currently operational, the Panguna mine was once the largest open pit copper mine in the world, producing tremendous revenue for Rio Tinto and Papua New Guinea. However, as a consequence of mine construction and production, thousands of acres of rainforest were cut down and billions of tons of mine waste were dumped into local rivers and the surrounding ocean, degrading drinking water quality and destroying fisheries and local fishing economies (Klare 2001; Langston 2004). Mine pollution may also have increased death rates on the island, especially among children (S. James 2006). Villagers living on or near the mine property were also forcibly removed from the area to make way for the mine (S. James 2006), and thousands of foreign miners were brought to the island, threatening traditional cultures and ways of life (Langston 2004).

When some Bougainvillians sought to rectify this situation by attacking the mine, closing it, and declaring independence from Papua New Guinea, the Papuan government responded by invading Bougainville (Klare 2001). Some people believe that the invasion was carried out at the behest of Rio Tinto, which may also have helped transport Papuan

troops to Bougainville during the conflict (Langston 2004). The resulting war left 15,000–20,000 dead (*Economist* 2008).

The recent history of Ivanhoe Mines Ltd. points again to the relationship between copper mining and armed violence. Ivanhoe began
investing in the Monywa copper mine in Myanmar in 1992 and was
soon a 50% owner of the mine.[19] Human rights organizations argue that
the infrastructure that surrounds and supports the mine was built by
forced labor (Cohn 2001), and Myanmar's military government, which
during the time period covered by this study used slave labor, brutally
repressed any opposition to its rule and routinely violated its citizens'
human rights,[20] had a large ownership stake in the mine (Hoffman 2007;
Stueck 2006). Since separating mining activities from armed violence
and the threat of armed violence in highly repressive regimes such as
Myanmar's is difficult to do, especially when the government is a major
mine owner, it is clear that copper mining is strongly associated with
armed violence in Myanmar.

Violence Associated with Other Minerals and Vital Natural Resources

The evidence presented in the copper and manganese case studies
and in table 5.2 supports my argument that natural resource extraction is often associated with armed violence. The evidence further
suggests that armed violence plays a key role in ensuring that natural
resources critical to industrial production and national security are
extracted and sold in sufficient quantities to promote capital accumulation and state power in the core (state power is, of course, predicated
on a strong national economy). Thus, armed violence appears to be
a key mechanism that helps make global capital accumulation, core
nation state power, the global ecological crisis, and ecological unequal
exchange possible.

The copper and manganese case studies also support my theoretical argument about why natural resource extraction is associated with
armed violence. As predicted, in four of the six case study countries—
Brazil, China, West Papua, and Bougainville—the national government
used armed violence to suppress anti-mine activists or insurgents who
were protesting or rebelling against the negative repercussions of min-

ing. In all four of these countries these negative repercussions included environmental degradation and the loss of livelihoods; in Brazil, West Papua, and Bougainville they included forced removal; in Brazil, China, and Bougainville they included health problems associated with or believed to be associated with mining; and in both West Papua and Bougainville the local residents were indigenous and colonized people who resented foreign rule (many of the local residents in Brazil were also indigenous people). Moreover, in the two case study countries where I did not find any evidence of anti-mine protests (Gabon and Myanmar), the governments were so highly repressive as to make protest or media coverage of protest highly unlikely. Finally, as I predicted, forced removal occurred in Brazil, West Papua, and Bougainville in order to make room for mining or mine-related activities.

Obviously, armed violence is not associated with the extraction of the entire supply of critical minerals examined in table 5.2 (or with the entire supply of all natural resources), and it is quite doubtful that the violent actions of the governments and companies examined in these tables were motivated by concern about core nation capital accumulation or the military power of wealthy nations. Nevertheless, by increasing the available mineral supply and keeping mines operational, armed violence likely helps to lower mineral prices and increase mining company profits, both of which support state power and capital accumulation in core nations, the former (low prices) by increasing corporate profits in general and the latter (mining company profits) by ensuring that mining companies are able to stay in business and extract the minerals that industrial economies and modern militaries require.

It is possible, of course, that the critical minerals examined in the case studies and table 5.2 are more strongly associated with armed violence than are other minerals and natural resources. To demonstrate that this is not the case, I now briefly discuss several examples of armed violence associated with the world's three largest mining companies (as of 2007) and with petroleum and rainforest timber extraction and transport.

In 2007 the world's three largest mining companies (in terms of total assets) were Rio Tinto, CVRD/Vale, and BHP Billiton. Rio Tinto, which has a major financial interest in the exceedingly violent Grasberg mine that I previously discussed (Rio Tinto 2004, 2013), also ran the now closed Kelian gold mine in Indonesia. In order to open this mine, Indonesia forcibly evicted hundreds of indigenous Dayak villagers from their land, and after the mine became operational, security guards at the mine raped, beat, and shot local community members in order to sow fear and prevent protest (Branford 2007). In the first few years of the 21st century Rio Tinto also relocated more than 900 people to make way for its Murowa diamond mine in Zimbabwe (Africa News Service 2003); and in 2007 it purchased the aluminum giant Alcan, which made it partners with the highly repressive Guinean government, which has violently suppressed mine worker strikes and anti-mine protests (BBC News 2008; Africa News 2007). Rio Tinto also benefited in the 1980s from the violent repression I discussed in Bougainville, Papua New Guinea (Branford 2007), and after 2005 invested heavily in Ivanhoe Mines, which operated in Myanmar until 2009 (see note 19) and benefited from that country's repressive military regime (Carlock 2010; Mines and Communities 2007).

CVRD/Vale also benefits from armed violence. It was the major corporate beneficiary of the violent Carajas Grande project that I discussed earlier in the chapter, and in 2006 it acquired the international mining giant Inco, which mined nickel in Guatemala for nearly 40 years on land expropriated by the authoritarian Guatemalan government from the indigenous Q'eqchi' people (Russell 2006). Inco also operated a nickel mine in New Caledonia, where authorities responded to a native islander anti-mine protest and blockade by deploying French troops at the blockade sites. The troops charged the blockades and at one point fired live ammunition at protesters (Mining Watch 2006).

By acquiring Inco, CVRD/Vale also took possession of the Sulawesi nickel mine in Indonesia. In order to make way for this mine, the authoritarian Suharto regime forcefully expelled the Karonsi'e Dongi people from the area, and the Karonsi'e Dongi have since suffered from widespread environmental degradation, a loss of livelihoods, and continued harassment and intimidation (Mikkelsen 2013; Mining Watch 2005). Finally, the development of CVRD/Vale's Moatize coal mine in Mozambique required the involuntary removal of approximately 5,000 people. Resettlement began in 2010, and in 2013 protestors claimed that they had not received sufficient compensation for their losses (Agencia de Informacao de Mocambique 2010; BBC News 2013; Keane 2012).

BHP Billiton has also relied on armed violence to commence and sustain its mining operations. For instance, force and violence were used to relocate people from their land in order to make way for the company's Tintaya copper mine in Peru and to protect the mine once it became operational (*Herald Sun* 2005); and in Colombia, military police forcefully evicted residents of the entire town of Tabaco to construct the world's largest coal mine, the Cerrejon coal mine, of which BHP Billiton is a major shareholder (Roberts and Trounson 2007; Solly 2010). Surrounding villages also suffered brutal attacks and massacres by paramilitary forces, though BHP Billiton denied involvement in these atrocities. BHP Billiton did, however, acknowledge that it had contracted with the Colombian military to provide protection to its employees (Robinson 2005).

Logging is also often associated with violence. Although much of this violence involves conflict between illegal loggers and indigenous people (*New York Times* 2007; Noble-Wilford 2008), state- and company-sponsored armed violence play a critical role in facilitating hardwood extraction in several nations, including Malaysia (Meo 2007), Indonesia (Allard 2005; Poulgrain 2005), Myanmar (MacKinnon 2007), Liberia (from 1989 to 2003; Bloomfield 2006), and the Democratic Republic of Congo (Vidal 2007). Armed violence has also played an important role in enabling the clearcutting of forests in Indonesia to make way for palm oil plantations designed to produce biodiesel (Green 2007; Knudson 2009). Violent actions associated with hardwood extraction and the clearing of forests for palm oil plantations include arrests, armed protection of loggers by soldiers, soldiers beating protestors, the wholesale demolition of villages, forced relocation, rape, and murder.

Armed violence also plays a critical role in facilitating petroleum and natural gas extraction and transport (Klare 2004). For example, in Nigeria oil production has sparked considerable armed violence and repression as local people, rebel groups, and Nigerian and oil company military personnel have struggled for control over the region's oil and local environment (J. Brown 2006; Human Rights Watch 1999; Watts 2008). One example of this occurred in 1990, when Nigeria's Mobile Police force responded to a protest at a Shell Oil Company facility by killing approximately 80 unarmed protestors and destroying hundreds of homes. Shell claims that it stopped requesting assistance from the Mobile Police after this incident, but a 1999 Human Rights Watch report disputes this claim.

Massive protests in 1993 against the Nigerian government and Nigerian oil production also resulted in state repression, which in this case included the widespread detention and harassment of activists as well as extrajudicial killings by military forces (Human Rights Watch 1999); and between 1993 and 1996 the Nigerian government, as part of its efforts to protect oil company interests, destroyed 27 villages, killed 2,000 people, and displaced 80,000 others (Kupfer 1996).

Armed violence and the threat of armed violence have also been used to ensure the safe extraction and transport of oil in Ecuador and natural gas in Myanmar. In Myanmar, security forces protecting the Yadana natural gas pipeline in the 1990s and first decade of the 21st century were accused of committing widespread human rights abuses, including "forced labor, land confiscation, forced relocation, rape, torture, [and] murder" (Earth Rights International 2011); and in Ecuador, documents from 2000 and 2001 indicate that the government agreed to spy on its citizens for Occidental Oil and to provide military protection for 16 oil companies that were operating within its borders (Hearn 2006).

Military violence and aid also play a critical role in U.S. efforts to exert control over global petroleum supplies (Klare 2004). For instance, in the first decade of the 21st century, the U.S. was engaged in two wars that were directly (Iraq) and indirectly (Afghanistan) related to petroleum production or transport, and in the 1980s the U.S. Navy protected Kuwaiti oil shipments from Iraqi attack in the Strait of Hormuz (Klare 2004).[21] The U.S. also provides military training and high levels of military aid to nations that either possess large oil reserves or, like Israel and Egypt, are located in regions with large oil reserves;[22] and the U.S. military command structure, including its new African Command (Africom), is designed in large part to protect and secure critical oil and mineral supplies (Berschinski 2007; Klare 2004).

It is thus quite clear that the use of armed violence to safeguard resource extraction and transport activities is not restricted solely to the critical minerals that I examined in the previous sections of the chapter. Instead, armed violence is associated with the extraction and transport of a wide range of natural resources that are critical to capital accumulation, core nation power, and the functioning of local, national, regional, and global economies.

Overlapping Mechanisms: The World Bank, Mining, and Violence

In the previous sections of this chapter I examined the association that exists between armed violence and natural resource extraction and transport, but I did so without demonstrating that armed violence works in conjunction with other elite-controlled mechanisms that also facilitate natural resource extraction and transport. I rectify this imbalance in my account by turning my attention in this section to the intimate association that exists in sub-Saharan Africa between World Bank sectoral adjustment programs, armed violence, and mining, all of which have serious human, social, and environmental consequences. I begin by discussing the effectiveness of World Bank–led mining sector reform in sub-Saharan Africa and then demonstrate that armed violence is strongly associated with sub-Saharan mining projects that have received World Bank funding and insurance underwriting.

Mining Sector Reform in Sub-Saharan Africa

Sub-Saharan Africa is one of the most structurally adjusted regions in the world. For instance, between fiscal years 1980 and 1996, 37 sub-Saharan nations received $15 billion worth of loans as part of 163 separate World Bank adjustment programs (World Bank 1997); and since 1979 roughly one-third of all World Bank lending in the region has gone to structural adjustment (World Bank n.d.).[23]

As described in detail in chapter 3, the goals of structural adjustment include opening up developing nations to foreign investment, drastically reducing government involvement in the economy, and increasing exports, particularly agricultural and natural resource exports, in order to increase developing nations' foreign cash reserves. Although structural adjustment programs are often designed for the general economy, many of these programs are directed toward specific economic sectors, such as the mining sector (this is true not just in Africa, but in Asia and Latin America too).

As a result, sub-Saharan African nations have not only undergone dramatic and widespread adjustments to their overall economies, many of them have also adopted extensive mining sector reforms (Campbell 2004; Chachage et al. 1993; Lambrechts 2009), with some observers claiming that 35 sub-Saharan countries had reformed their mining codes by 1996, presumably in response to World Bank pressure (UNCTAD 2005). Consistent with this trend, 30 African nations crafted new mining acts between 1990 and 2000, with 12 African nations "reviewing their act [in 2000], and only 13 [national] acts predat[ing] 1990" (Otto et al. 2006, xvi).

As part of its efforts to get African nations to reform their mining codes, the World Bank routinely uses mining sector reforms undertaken by one African nation to convince other African nations that if they want to remain internationally competitive, they must reform or further reform their own mining codes, with new codes giving mining companies greater advantages and African countries fewer advantages than did earlier mining codes. This, of course, fosters a race to the bottom in which reforms increasingly diminish the ability of African nations to defend their mineral rights and profit from their natural resources, thus greatly benefiting multinational mining corporations (Akabzaa 2004). Ghana's mining sector reforms were, for example, held up as a model to Tanzania, Mali, Zambia, Burkina Faso, Guinea, and the Democratic Republic of Congo, whose later, more far-reaching mining sector reforms were then held up as the new ideal toward which Ghana and other early reformers should strive (Akabzaa 2004).

Although it is difficult to accurately determine which sub-Saharan nations have reformed their mining codes in response to World Bank pressure, it is fairly clear that Guinea, Liberia, Zambia, Burkina Faso, Tanzania, Senegal, the Democratic Republic of Congo, Uganda, Burundi, Ghana, Mali, Madagascar, Sierra Leone, Cameroon, Central African Republic, and Mauritania have done so;[24] and judging from World Bank statements, it is likely that other sub-Saharan nations have done so as well (*Mining Journal* 2000).

The main goal of mining sector reform has been to improve the performance of the mining sector by altering the economic and political incentives that help determine whether mining companies decide to invest and operate in a country (Campbell 2004), with performance measured in terms of factors such as production volume, foreign investment in the sector, and the value of mineral exports. Because increased foreign investment has generally been viewed as the catalyst for increasing production volume and exports, reforms have generally been designed to decrease the role of the state and increase the role of private business in the sector, to decrease the tax, labor, and other financial costs of doing business in the sector, and to otherwise make the investment climate more friendly and stable for foreign mining companies (Campbell 2004). Mining reforms thus include measures such as privatizing state-owned mines, shifting the role of the state so that it no longer runs the sector but only regulates it, allowing mining companies to more easily repatriate profits, drastically reducing the taxes mining companies pay, shifting these taxes from royalties based on production levels to taxes based on profit levels, making it easier for mining companies to move from mineral exploration to mineral exploitation, and allowing mining companies to easily transfer their mining rights to other privately held companies (Campbell 2004; Lambrechts 2009; Naito et al. 2001; UNCTAD 2005).

Sub-Saharan nations that have undertaken these reforms have experienced dramatic increases in mining sector investment, mining sector privatization, and mineral exports. For instance, between 1994 and 1998 $1.39 billion of sub-Saharan African mining assets were privatized (Pigato 2000); in 2004 $15 billion was invested in the African mining sector, representing 15% of the world total, or three times the percentage the continent received in the mid-1980s (UNCTAD 2005); and in Ghana, mining reforms begun in the mid-1980s resulted in dramatic increases by 2004 in the "annual production of gold (500 percent), diamonds (300 percent) and bauxite (more than 600 percent)" (Hilson 2004, 54). Mineral exports also increased rapidly in Tanzania after that country enacted mining sector reforms, with diamond production increasing by 1,390% and gold production by 400% between 1994 and 2000 (Akabzaa 2004).

Despite all this, sub-Saharan Africa has not benefited appreciably from mining sector reform. There are four main reasons for this. First, the World Bank and IMF generally insist that structurally adjusted nations use their export revenues to pay off their debts rather than to invest in their economies and infrastructure. Second, because the goal of mining sector reform is to increase mineral exports rather than to use or process these minerals domestically, the mining sector has not been integrated into the broader economies of those sub-Saharan African nations that have undergone reform. As a result, mining has not acted as an engine for employment and growth in other sectors of these nations' economies (DanWatch and Concord Danmark 2010; Lambrechts 2009; UNCTAD 2005).

Third, in mineral-rich sub-Saharan African nations the mining sector represents only a very small portion of national gross domestic product (GDP). For instance, in 2002 mining represented only 2% of Tanzania's GDP, even though minerals brought in more than one-third of that nation's export revenue (UNCTAD 2005); and in the late 1990s and 2000 mining contributed only 2% to the GDP of Ghana, one of Africa's largest mineral exporters, despite being that nation's largest foreign exchange earner since 1992 (Campbell 2004).[25] In fact, according to Steve Manteaw, a leading expert on the Ghanaian mining industry, once the environmental costs of mining are taken into account, the sector actually has a negative impact on Ghana's GDP (DanWatch and Concord Danmark 2010).

Fourth, due to several factors, including the low tax and royalty rates established by industry friendly mining sector reforms, the ability of many mining companies to negotiate deals with sub-Saharan African nations to pay even lower tax and royalty rates than those set out in mining sector legislation, and the ability of mining companies to legally and illegally move much of their earnings offshore, sub-Saharan African nations often receive much less mining revenue than one would expect given the value of their mineral exports (Campbell 2004; DanWatch and Concord Danmark 2010; Lambrechts 2009). For example, between 2002 and 2006 the Tanzanian government only received $87 million in royalties on $2.9 billion in mining company revenue (Lambrechts 2009); in Zambia in 2006–2007 Konkola Copper Mines only paid a 0.6% royalty rate ($6.1 million) on more than $1 billion in revenue (Dymond 2007; Lambrechts 2009); and between 1992 and 2004 tax revenues and other remittances that copper mining companies paid Zambia declined from about $200 million to $8 million despite the fact that (a) Zambia produced roughly the same amount of copper in 2004 as in 1992 and (b) copper prices were about 25% greater in 2004 than in 1992 (Lambrechts 2009). It should therefore come as no surprise that mining contributes very little to the economies of most sub-Saharan African nations or that between 2002 and 2006 dramatic increases in mineral prices and mining company profits did not translate into greater mining revenues for mineral-rich African nations (DanWatch and Concord Danmark 2010).

Armed Violence and World Bank–Supported Mining Projects

In addition to contributing little to economic development in sub-Saharan Africa, mining in the region is also strongly associated with armed violence. To demonstrate this association, tables 5.3a and 5.3b present evidence that Eric Bonds, Katherine Clark, and I collected on 21 sub-Saharan African mining projects that received funding and/or insurance underwriting from the World Bank's International Finance Corporation (IFC) or Multilateral Investment Guarantee Agency (MIGA) at some point between 1997, the first year for which we had complete data on IFC lending, and 2009, the year in which we conducted the research. We determined which mining projects to include in the study by selecting from a database of all IFC- and MIGA-supported projects those sub-Saharan African mining projects in which mining, mine construction, or mineral exploration had commenced by the time we began conducting our research.[26]

Given this data collection strategy, five of the 21 mining projects listed in tables 5.3a and 5.3b had not yet reached the construction stage when we conducted our research. These projects are kept separate from the other projects in the two tables since armed violence is less likely to occur during the exploration stage of a mining project than during the mining and mine construction stages.

Table 5.3a lists the 21 mining projects in alphabetical order, with the five exploration projects listed in a separate section at the end of the table. For each project, the table indicates which nation hosted the project; whether World Bank–supported mining reforms had been enacted in that nation by 2009 and, if so, whether these reforms were enacted before, during, or after World Bank involvement in the project;[27] the main minerals produced or expected to be produced by the project; the years and degree of IFC and MIGA involvement in the project; and the Political Terror Scale (PTS) scores of the nation hosting the project, derived from Amnesty International and U.S. State Department data. It is important to note that the PTS scores in table 5.3a represent the average

TABLE 5.3A. The World Bank, Mining, and Violence in Sub-Saharan Africa

Mine	Nation	Mining reform	Mineral	IFC investment	MIGA underwriting	PTS[a]
Ahafo Ghana	Ghana	Early 1980s (B)[b]	Gold	2006–ongoing[c] (loan)	None	SD: 3 AI: 2
Anvil Mining	Congo (DRC)	2002 (B)	Copper and silver	None	2004–2009 ($13.6m)	SD: 4.5 AI: 5
Chambishi Metals Plc	Zambia	Mid-1990s (B)	Refining cobalt from copper	None	2000–2003 ($30m)	SD: 2.75 AI: 3
Essakane Gold	Burkina Faso	1997 (B)	Gold	None	2008–2009 ($190m)	SD: 3 AI: 3
Ghanaian-Australian Goldfields Limited	Ghana	Early 1980s (B)	Gold	1990–2007 (20% owner of mine)	1993–1996	SD: 1.7 AI: 2.3
Hernic Ferrochrome	South Africa	Do not know	Ferrochrome	2004–ongoing (5% owner of mine)	None	SD: 3.2 AI: 3
Kahama Mining Corp. (Bulyanhulu Mine)	Tanzania	1995 (B)	Gold	None	2000–2003 ($56.25m)	SD: 3 AI: 3
Kasese Cobalt	Uganda	2004 (A)	Processing copper mine slag/waste	1997–2002 (8% owner of mine)	1993–2002 ($63m)	SD: 3.2 AI: 3.5
Kenmare Moma	Mozambique	2001 (B)	Titanium and zircon	None	2003–ongoing ($32.5m)	SD: 3 AI: 2.7

Mine	Nation	Mining reform	Mineral	IFC investment	MIGA underwriting	PTS[a]
Kolwezi Tailings Project	Congo (DRC)	2002 (B)	Cobalt and copper	2005–ongoing (7%–10% owner of mine)	None	SD: 4.5 AI: 5
Konkola Copper	Zambia	Mid-1990s (B)	Copper	2000–2002 (7.5% owner of mine)	None	SD: 2.7 AI: 3
Lonmin	South Africa	Do not know	Platinum group metals	2007–ongoing (loan and $15m–$50m in company equity)	None	SD: 3.5 AI: 3
Magadi Soda	Kenya	Do not know	Soda ash	1994 and 2005 (loans)	None	SD: 4 AI: 3.5
Mvelaphanda Gold	South Africa	Do not know	Gold	2003–2009 ($27m investment in company)	None	SD: 3.2 AI: 3
Sadiolia Gold	Mali	1991 or 1993 (B)	Gold	1994–2009 (6% owner of mine)	1996–2000 ($50m)	SD: 1.9 AI: N/A
Syama Gold	Mali	1991 or 1993 (D)	Gold	1991–2004 (loan and 15% owner of mine)	None	SD: 2 AI: 2.4
Projects without an existing mine at the time of the study						
AEF Fasomine	Burkina Faso	1997 (B)	Gold	1998–2000 (10% owner of mine)	None	SD: 2 AI: 2.3

TABLE 5.3A. The World Bank, Mining, and Violence in Sub-Saharan Africa (*cont.*)

Mine	Nation	Mining reform	Mineral	IFC investment	MIGA underwriting	PTS[a]
Baobab Resources	Mozambique	2001 (B)	Iron, gold, copper, silver, PGM,[d] diamonds	2008–ongoing (11% owner of company, 15% owner of mine)	None	SD: 3 AI: 3
Kalahari Diamond	Botswana	Do not know	Diamonds	2003–2005 (9.4% owner of mine)	None	SD: 2 AI: N/A
Kalukundi Cobalt-Copper	Congo (DRC)	2002 (B)	Cobalt and copper	2007–ongoing ($8m in company shares)	None	SD: 5 AI: 5
Simandou II	Guinea	1995 (B)	Iron	2007–ongoing (5% owner of mine)	None	SD: 3.5 AI: 4

[a] PTS: Political Terror Scale score; SD: State Department; AI: Amnesty International.
[b] The letters B, D, and A indicate whether reforms were enacted before (B), during (D), or after (A) World Bank involvement in the mining project.
[c] Throughout the table, "ongoing" means ongoing as of 2009, when the data for this table were collected.
[d] PGM: Platinum group metals.

scores for the host country during the years of World Bank involvement in the project and thus can differ for different projects hosted by the same country. In addition, please recall that the level of terror increases as the PTS score increases and that a score of 3 indicates that "there is extensive political imprisonment, or a recent history of such imprisonment [in the country]," that "execution or other political murders and brutality may be common," and that "unlimited detention, with or without a trial, for political views is accepted" (Gibney et al. 2009).

It is also important to note that the evidence presented in tables 5.3a and 5.3b was gathered using the same newspaper and archival search strategies as were employed to gather the evidence presented in tables 5.1 and 5.2. As a result, tables 5.3a and 5.3b likely underrepresent the degree to which violence is associated with IFC- and MIGA-supported mining projects in sub-Saharan Africa.

Table 5.3a shows that Botswana, Guinea, Kenya, Tanzania, and Uganda each hosted one of the 21 World Bank–supported projects, that Burkina Faso, Ghana, Mali, Mozambique, and Zambia each hosted two of these projects, and that the Democratic Republic of the Congo (DRC) and South Africa each hosted three of these projects. Sixteen of the 21 projects were hosted by nations that had enacted World Bank–supported mining sector reforms by 2009, with 14 projects hosted by nations that enacted mining reforms before World Bank involvement in the mining project, one project hosted by a nation that enacted mining reforms during World Bank involvement in the project, and one project hosted by a nation that enacted mining reforms after World Bank involvement in the project. Average PTS scores of 3 or greater (based on either Amnesty International or State Department data) were recorded for nations hosting 16 of the mining projects, with all three projects in the DRC receiving a PTS score of 5, indicating that when the World Bank funded or underwrote the DRC projects, "terror ha[d] expanded to the whole population, . . . [and the nation's] leaders . . . place[d] no limits on the means or thoroughness with which they pursue[d] personal or ideological goals" (Gibney et al. 2009).

Finally, the fifth column in table 5.3a (IFC investment) indicates that in addition to supporting the 21 mining projects, the World Bank was actually a partial owner of 11 of them (for example, the entry for Ghanaian-Australian Goldmines Limited indicates that until 2007 the

World Bank had a 20% ownership stake in the mining project). In four of the 21 projects, the World Bank also partially owned one of the mining companies involved in the project (in only one of these four projects did the World Bank own shares in both the company and the mine). Thus, any of the profits generated by these projects and mining companies, and any of the violence employed to make these projects and mining companies profitable, directly benefited the World Bank.

To determine the degree to which armed violence is associated with the 21 mining projects, table 5.3b provides detailed information on political and human rights in the countries hosting the mining projects (the host countries) and on instances of armed violence that can be linked to the 21 mining projects, to other mining projects located in the host countries, and to the mining companies involved in the 21 projects. To include as much information in the table as possible, acronyms are used to denote regularly occurring table entries. The definitions of these acronyms are listed at the end of the table.

The data on political and human rights are drawn primarily from U.S. State Department reports on human rights in the host nations in the year prior to World Bank involvement in each project, providing the reader with a sense of the human and political rights situation in these countries when the World Bank was deciding whether to support the projects (supplemental information on human and political rights was obtained from Amnesty International reports, other human rights reports, and news articles). For five of the mining projects we had to use a State Department report written for a different year than the one we wanted to use. This is clearly indicated in the table. The remaining evidence in table 5.3b is drawn from articles and reports that Katherine Clark, Eric Bonds, and I obtained when undertaking our newspaper and archival research.

Table 5.3b also indicates (in parentheses) whether the violent acts recorded in the table occurred before, during, or after World Bank involvement in each of the projects. This timing is important since the World Bank, with its vast information gathering abilities, presumably knew about (or could have found out about) violent activities that occurred before its involvement with these companies and projects and thus could have avoided these companies and projects if it so desired. Conversely, violent acts that occurred during or after its involvement in

TABLE 5.3B. World Bank, Mining, and Violence in Sub-Saharan Africa

Mine	Political and human rights abuses	Police, military, mine security, or rebel violence specific to project	Forced removal	Mine-related armed violence elsewhere in the country	Mining companies associated with violence outside the country
Ahafo Ghana (Ghana)	K, B, DA, AA, JIBH, BRP, forced CL, T, LFA: 2005 Report	Beatings, arrests (during); fired at protestors, killing 1 and injuring others (before)	10,000 (during) and another 10,000 possible	B, K, FR, MS (during)	Yes (before and during)
Anvil Mining (DRC)	K, T, B, R, AA, DA, LFP, LFS, LFA, LRCG, GIEV, CL, RKTRC, BRP, MS, JIBH, AR, FNPR: 2003 Report	Troops put down local rebellion with transportation provided by Anvil mining; troops beat, torture, and kill up to 100 people (during)	None	War and violence by multiple nations and rebels to secure mineral resources (before and during)	Yes (during)
Chambishi Metals Plc (Zambia)	K, T, B, AA, DA, LFP, ARPO, LRCG: 1999 Report	Police provide mine security during a 5-day strike	None	VU, K, firing at protestors; police called in to put down miner uprisings (after)	Yes (during)
Essakane Gold (Burkina Faso)	B, T, AA, DA, LFP, LFA, LFS, CL, BRP, JIBH: 2007 Report	None	11,545 (during); 1,000 economically displaced (during)	None	Yes (before and during)
Ghanaian-Australian Goldmines Limited (Ghana)	K, B, AA, DA, JIBH, BRP, forced CL, LFP: Should be 1989 Report but used 1999 Report	Beatings, arrests, torture, and use of violent dogs; people crossing mine property shot at (during)	20,000 (before); 173 families (during)	B, K, FR, MS (during)	Yes (during)

211

TABLE 5.3B. World Bank, Mining, and Violence in Sub-Saharan Africa (*cont.*)

Mine	Political and human rights abuses	Police, military, mine security, or rebel violence specific to project	Forced removal	Mine-related armed violence elsewhere in the country	Mining companies associated with violence outside the country
Hernic Ferrochrome (South Africa)	B, T, K, R, DA, CL, LFP: 2003 Report	None	None	Police beat, teargas, and fire live and rubber bullets into crowds protesting mine expansion (during)	No
Kahama Mining Corp. (Bulyanhulu Mine) (Tanzania)	K, B, T, AA, LFP, LFS, LFA, ARPO, CL, CLM, BRP: 1999 Report	Residents and NGOs claim that 52 artisanal miners were buried alive (before); lawyers arrested and evidence they had regarding the 52 miners and forced removal was confiscated (during)	NGOs, relying on local police reports, claim 200,000 forcibly removed (before)	FR (during); K and arrests (during and after); in 2003 the government gave African Gem Resources permission to fire live bullets on any artisanal miner (during)	Yes (during and after)
Kasese Cobalt (Uganda)	K, T, B, AA, DA, LRCG, LFA, CL, BRP, LFP, LFS, FNPR, GIEV: Should be 1992 Report but used 1999 Report	Rebels kill civilians at the mine in an attack (during)	None	None	Yes (during and after)
Kenmare Moma (Mozambique)	K, T, AA, DA, B, LFA, BRP, CL: 2002 Report	None	Possibly 1,000	FR (before and after); K and fire at crowd (during); police arrest thousands of unpermitted miners (after)	No

Mine	Political and human rights abuses	Police, military, mine security, or rebel violence specific to project	Forced removal	Mine-related armed violence elsewhere in the country	Mining companies associated with violence outside the country
Kolwezi Tailings Project (Kingamyambo Musonoi) (DRC)	K, T, B, R, M, DA, LRCG, LFP, LFS, LFA, CL, RKTRC (including child soldiers), AA, FNPR, BRP, GIEV, JIBH, forced labor, CLM, MS: 2004 Report	Mining companies provide resources and offer almost $800 million to various belligerents in the Congolese war to secure access to minerals (before)	76 plus economic displacement (during)	War and violence by multiple nations and rebels to secure mineral resources (before and during)	Yes (before)
Konkola Copper (Zambia)	K, B, DA, AA, T, LFP, LRCG, LFA, CL, LFA, BRP, VU (protesting miners were shot by police): 1999 Report	Police use force to put down miner strike and community protests and to expel artisanal miners from property; police fire at miners, wounding many and killing 1 (after)	750 (during)	VU, K, police called in to put down miner uprisings (after)	Yes (during and after)
Lonmin (South Africa)	K, B, R, T, DA, BRP, LFP, CL: 2006 Report	None	None	Police beat, teargas, and fire live and rubber bullets into crowds protesting mine expansion (during)	No
Magadi Soda (Kenya)	K, B, T, R, DA, AA, BRP, LFP, LFS, LFA, JIBH: Should be 1993 report but used 1999	23 arrested after mine protest (before second investment); beatings and arrests at another protest (during)	None	FR, and police disrupt NGO's ability to assist affected persons (during)	Yes (before, during, and after)

TABLE 5.3B. World Bank, Mining, and Violence in Sub-Saharan Africa (*cont.*)

Mine	Political and human rights abuses	Police, military, mine security, or rebel violence specific to project	Forced removal	Mine-related armed violence elsewhere in the country	Mining companies associated with violence outside the country
Mvelaphanda Gold (South Africa)	K, B, T, R, DA, BRP, forced CL, LFS, LFP: 2002 report	None	None	Police beat, teargas, and fire live and rubber bullets into crowds protesting mine expansion (during)	Yes (before and during)
Sadiolia Gold (Mali)	LFA, forced CL (many more abuses in previous years): Should be 1993 report but used 1999	None	1,115 plus 3 villages (during)	None	Yes (during and after)
Syama Gold (Mali)	LFA, forced CL (many more abuses in previous years): Should be 1990 report but used 1999	None	Residents possibly lost access to communal agricultural lands	Possible FR	Yes (during and after)
Projects without an existing mine at the time of the study					
AEF Fasomine (Burkina Faso)	LRCG, K, AA, DA, T, BRP, LFP, LFS: Should be 1997 report but used 1999	None	None	None	No
Baobab Resources (Mozambique)	K, AA, DA, LFP, B, T, LFS, JIBH, LFA: 2007 report	None	None	FR (before and during); K and fire into crowd (before); police arrest thousands of unpermitted miners (before)	No

214

Mine	Political and human rights abuses	Police, military, mine security, or rebel violence specific to project	Forced removal	Mine-related armed violence elsewhere in the country	Mining companies associated with violence outside the country
Kalahari Diamond (Botswana)	B, DA, LFP: 2002 report	Police arrest, beat, and torture indigenous San Bushmen (before, during, and after)	2,000 (before)	None	Yes (before, during, and after)
Kalukundi Cobalt-Copper (DRC)	K, T, R, AA, DA, RKTRC, LFP, LFA, BRP, forced CL (including child soldiers and in mines), JIBH, GIEV: 2006 report	None	2,000 (expected)	War and violence by multiple nations and rebels to secure mineral resources (before and during)	Yes (before and during)
Simandou II (Guinea)	K, B, T, R, DA, LRCG, AA, LFP, LFS, LFA, CL, BRP, JIBH, FNPR: 2006 report	None	None	B, K, R, T, arrests (during); K and fired at protestors (before and during)	Yes (before and during)

Notes: K, T, B, R, and M refer to state sponsored killing, torture, beatings, rape, and massacres, respectively; AA: arbitrary arrest; DA: detainee abuse; AR: authoritarian regime; LFP: limited freedom of the press; LFS: limited freedom of speech; RKTRC: regular killing, torture, rape, and conscription of citizens; FNPR: few to no political rights; LRCG: limited right to change government; GIEV: government instigated ethnic violence; ARPO: arrest reporters and political opposition; BRP: brutally repress protests; FR: forced removal; MS: state provides mine security; VU: violent uprising against mining; JIBH: journalists intimidated, beaten, or harassed; CL: child labor; CLM: child labor in mines.

these projects likely helped the World Bank meet its goal of increasing developing nations' natural resource exports.

Turning our attention to the second column in table 5.3b, we see that human rights abuses were fairly widespread in the nations hosting the 21 mining projects. These abuses included arbitrary arrests, detainee abuse, and state sponsored beatings, killings, torture, rapes, and massacres. Many of the nations hosting these projects also provided their citizens with few to no political rights and with limited freedom of speech and press.

Violence was also directly associated with many of these projects. Thus, columns 3 and 4 show that armed violence or threatened violence by police, military, mine security, or rebel forces occurred at 12 of the 16 projects that had advanced beyond the exploration stage and at two of the five projects that had not advanced beyond this stage. Violent acts associated with these mining projects included beatings, arrests, forced removal, torture, and killings (including 52 killings at one mine and up to 100 at another). They also included police providing mine security during a strike, police and military forces using violent tactics to put down strikes, and rebels and the DRC government fighting for control over mineral resources in the region where the Kalukundi Cobalt-Copper project is located (this last example is actually taken from column 5). In one case (Kasese Cobalt) rebels attacked a mine that was the largest source of foreign direct investment in Uganda, and in the case of the Kolwezi Tailings Project in the DRC, mining companies provided resources, including offers of nearly $800 million, to wartime belligerents who controlled or were fighting over access to the nation's vital mineral resources, including the cobalt and copper tailings at the Kolwezi Tailings Project. Forced removal also occurred at 10 project sites (column 4), with between 10,000 and 20,000 individuals forced to leave their homes at three sites and more than 200,000 people supposedly forced to move from the Bulyanhulu mining concession in Tanzania. Moreover, most of the violence highlighted in columns 3 and 4 occurred before or during the World Bank's involvement in the projects.

Fifteen of the 21 projects were also located in countries in which armed violence was used to facilitate resource extraction activities at mines not listed in the table (column 5), and 16 of the projects were at least partially owned by a company (not including the World Bank) that operated a mine that was associated with armed violence, often extremely high levels of armed violence, outside the host country (column 6). For example, Anglo American, a partial owner of Ghanaian-Australian Gold Mines Limited, was linked to violence and extensive human rights violations in Colombia, the Philippines, and South Africa during the World Bank's involvement in the Ghanaian-Australian project; Rio Tinto, which was involved in the Simandou II mining project, was linked to violence and extensive human rights violations in Indonesia both before and during the World Bank's involvement in the Simandou II project; and Newmont Mining, which operated the Ahafo Ghana mine, was implicated in extensive human rights violations at its Yanacocha gold mine in Peru (where police and private security forces have beaten, teargassed, and likely killed anti-mine activists) both before and during the World Bank's involvement in the Ahafo project (the World Bank also partially owns, or owned, the Yanacocha mine).

It is thus clear that in sub-Saharan Africa the World Bank is involved as a lender, underwriter, and/or owner in many mining projects that are associated with armed violence, either at the project site, in the host country, or at mining sites outside the host country that are owned or partially owned by a company involved in one of the 21 mining projects investigated here. In fact, there is only one project included in tables 5.3a and 5.3b—AEF Fasomine in Burkina Faso—that has no entries listed in the last four columns of table 5.3b, and this is only because another Burkina Faso project (Essakane Gold), which *is* linked to armed violence, is also listed in the table. Moreover, as column 2 of table 5.3b shows, during the World Bank's involvement in the Fasomine project, Burkina Faso's government abused prisoners, committed extrajudicial killings and arbitrary arrests, and limited the freedom of the press (U.S. Department of State 1999).

Conclusion

The evidence presented in this chapter demonstrates that armed violence is associated with the extraction of many critical and noncritical natural resources, suggesting quite strongly that the natural resource base on which industrial societies stand is constructed in large part through the use and threatened use of armed violence. Consistent with the theoretical argument I set forth at the beginning of this chapter, the evidence also demonstrates that when armed violence is used to protect resource extraction activities, it is often employed in response to popular protest or rebellion against these activities. Also consistent with my theoretical argument, the evidence shows that World Bank–supported mining projects in sub-Saharan Africa are strongly linked to armed violence and that structural adjustment, sectoral adjustment, and other World Bank pressure to reform national mining codes and institutions have helped to open sub-Saharan Africa up to multinational mining companies and projects, many of which are closely linked to armed violence. It thus appears that although the World Bank views the reform of government institutions and regulations as a necessary requirement for the profitable extraction of sub-Saharan Africa's mineral resources, in many instances this profitable extraction also requires the use of armed violence.

These findings, and the theoretical argument I set forth at the beginning of the chapter, extend prior sociological thinking and research on the environment in several important ways. First, very few environmental sociologists have examined armed violence and militarism, and those who have, have generally restricted their attention to military inequality between nations, military violence between nations, and the direct environmental consequences of weapons production, military activity, and war. Thus, this chapter establishes more clearly than prior environmental sociological theory and research the degree to which armed violence underpins the global ecological crisis. Second, in identifying armed violence as an important mechanism promoting ecological unequal exchange, this chapter contributes to the ecological unequal exchange literature, which has focused more attention on establishing the existence of ecological unequal exchange than on identifying the mechanisms responsible for it.[28]

Third, macro-structural environmental sociologists rarely discuss the intimate and pervasive link that exists between organizational and institutional violence on the one hand and local and global environmental degradation on the other hand. Thus, in demonstrating in this and the preceding chapters (as well as in the following chapter) that the organizations, institutions, and networks within which elite-controlled mechanisms are housed tend to be violent in at least one of three ways—they either (a) use or threaten to use armed violence to achieve their goals, (b) achieve their goals by relying at least in part on armed violence carried out by other organizations, institutions, or networks, or (c) severely harm individuals, communities, societies, and/or the natural world (*unarmed violence*)—I have significantly expanded environmental sociology's theoretical, conceptual, and empirical purview.

Fourth, in addition to demonstrating that much of the armed violence associated with natural resource extraction is carried out by developing nation governments, mercenaries, and rebels, I have also set forth a theoretical argument that explains why developing nations are likely to use armed violence to achieve their resource extraction goals even when doing so promotes ecological unequal exchange and the continued domination of these nations by core nations. Specifically, I have argued that developing nations often use armed violence to protect resource extraction activities not simply because these activities provide government leaders with money, wealth, and power, but more importantly because a set of overlapping organizational, institutional, and network-based mechanisms controlled by core nation governments and elites force developing nations to support these activities even when these activities enrich core nations and their elites at the expense of (or more than) developing nations and their elites. Thus, this chapter provides an explanation for why developing nations sometimes use armed violence to achieve resources extraction goals that contradict their long-term interests and suggests that core nations and corporations are able to distance themselves from many violent actions that benefit them, actions that they might otherwise have to take themselves.

In addition to the foregoing, this chapter also provides strong support for the IDE model. The most important way in which it does this is by demonstrating that armed violence is, in fact, one of several overlapping organizational, institutional, and network-based mechanisms that provide core nation elites with the means to control or gain disproportionate access to the natural resource wealth of developing nations, thereby promoting capital accumulation and military power in the core, degrading local, regional, and global environments, producing human suffering, and creating the conditions within which ecological unequal exchange can occur.

In this regard, it is important to note that, as established in chapter 3, the World Bank is essentially run by the U.S. and other powerful Western nations. World Bank structural adjustment policies, which
have forced developing nations around the world to increase their agricultural, mineral, and other natural resource exports, were also largely
dictated to the World Bank by the U.S. in the 1980s at the behest of
U.S. economic elites (see chapter 3) and have been strongly shaped by
the U.S. ever since. This implies, of course, that the organizational, institutional, and network-based mechanisms that give core nation governments and elites the power to shape World Bank policy likely also
allowed these governments and elites to shape the World Bank's sub-
Saharan mining policy. It further implies that the overlapping mechanisms that core nation elites rely on to ensure the safe and profitable
extraction of sub-Saharan African minerals are not restricted solely to
armed violence, mining reform, and structural and sectoral adjustment
but also include the organizational, institutional, and network-based
mechanisms that core nation elites use to shape their own countries'
foreign and international finance policies and the World Bank's lending
policies. Moreover, because armed violence and World Bank–imposed
structural adjustment, sectoral adjustment, and mining sector reform
are associated with resource extraction activities around the world, these
mechanisms play an important role in promoting resource extraction
not just in sub-Saharan Africa but globally as well.

It thus appears that to effectively influence social, economic, political, and military processes and outcomes around the world, core nation elites employ a wide range of organizational, institutional, and
network-based mechanisms that are nested within each other, operate
and interact at multiple geographic levels, and severely harm individuals, societies, and the environment. As a result, we cannot understand
or solve the global ecological crisis by focusing our attention primarily on core nation consumers, individual behavior, or one or two elite-
controlled mechanisms, but instead must trace the flow of elite power
back from the actual physical manifestations of the environmental crisis
to the multiple organizational, institutional, and network-based mechanisms that interact both to create the crisis and to form the global social
structures that make capital accumulation possible. Only in this way
can we identify the organizational, institutional, and network-based

mechanisms that are primarily responsible for the world's most severe social and environmental problems and trace responsibility for these problems back to the national and global elites whose decisions ultimately created them.

In addition to providing overall support for the IDE model and highlighting the importance of nested, interacting, and geographically dispersed elite-controlled mechanisms, the evidence presented in this chapter also strongly supports five of the six IDE hypotheses set forth in chapter 2, in particular the hypotheses that hold that organizational, institutional, and network-based inequality (OINB inequality) provide elites with the means to (a) monopolize decision making power, (b) shift environmental and non-environmental costs onto others, (c) inhibit the development and/or dissemination of environmental knowledge, attitudes, values, and beliefs, (d) divert public attention away from what elites are doing so that their actions will not be scrutinized, questioned, or challenged, and (e) frame what is and is not considered to be pro-environmental behavior, policy, and development.

The first of these five hypotheses (monopolized decision making power) is supported by three pieces of evidence. First, because the resource rich developing nations highlighted in this chapter are generally governed by authoritarian regimes that limit their citizens' political rights, these nations' mining decisions, including their decisions to use armed violence to support mine-related activities, are made by a small group of political and economic elites who can disregard the wishes of the citizenry. Second, armed violence is a form of organizational power that is controlled in any nation by a small group of military, political, police, and, in some cases, rebel elites who generally make all the key decisions regarding its use without any input from the general population. Third, regardless of whether developing nations are democratic or authoritarian, the World Bank is a highly undemocratic institution that has imposed a specific type of mining sector reform and mineral development on developing nations around the world irrespective of these nations' interests or the wishes of their leaders and citizens. It is thus quite clear that the overlapping mechanisms described in this chapter provide a small group of elites in core and periphery nations with the means to monopolize decision making power regarding mining and the use of armed violence.

Consistent with my second hypothesis, the organizational and institutional mechanisms highlighted in this chapter also provide core nation elites with the power to shift many of the environmental and non-environmental costs of mining away from themselves and most core nation citizens and onto citizens of the developing world. To some degree, shifting these costs onto the developing world also helps to shape the environmental knowledge, attitudes, values, and beliefs of core nation citizens and to divert these citizens' attention away from what core nation elites are doing by making the human, social, and environmental costs of industrial production and consumption much less visible to them. This, of course, undermines the argument, presented in chapter 1, that consumers possess the information they need to shape the social and environmental practices of business. Conversely, it is consistent with my hypotheses that hold that OINB inequality promotes environmental degradation by providing elites with the means to (a) shape citizens' environmental knowledge, attitudes, values, and beliefs and (b) divert citizens' attention away from what elites are doing.

Moreover, in the case of mining, the ability of core nation elites to divert citizens' attention away from what they are doing and mold citizens' environmental understandings and conceptions is achieved in a much more direct and effective manner by the mainstream news media, an important organizational mechanism that is also controlled by economic elites.[29] For instance, as demonstrated in chapter 3, the *New York Times* provides readers with virtually no information on the role the U.S. plays in shaping World Bank policy or on the role structural adjustment plays in shaping developing nation policy; nor does it provide more than the barest of information on the environmental and social consequences of World Bank loans, including loans for mineral exploration and mining reform. And though my colleagues and I relied on the *New York Times* and other mainstream news sources to generate much of the evidence presented in this chapter, it took hundreds of hours to gather this evidence, many of the news sources used in the chapter are published outside the U.S., and much of the evidence employed in the chapter comes from NGO and United Nations reports. In addition, my colleagues and I never came across a mainstream news article that synthesized the relevant information in such a way as to make the World Bank's role in shaping developing nation mining policy, the United

States' role in shaping World Bank policy, or the widespread social and environmental implications of mining fully apparent to readers. The mainstream news media in the United States thus fails to provide U.S. citizens with the information they need to fully evaluate World Bank policy regarding mining in developing nations or to fully understand the human, social, and environmental costs of mining, the World Bank's and United States' role in imposing these costs on developing nations, or the degree to which the U.S. influences World Bank and developing nation policy. This not only undermines democracy in the U.S. by failing to provide U.S. citizens with the information they need to effectively influence their government's policies, it also shapes U.S. citizens' environmental knowledge, attitudes, values, and beliefs by diverting their attention away from the social and environmental consequences of what the World Bank and U.S. and global elites are doing.

U.S. citizens' attention is further diverted from what the World Bank and U.S. and global elites are doing because much of the armed violence associated with natural resource extraction is carried out by developing nation governments, mercenaries, and rebels. This, as I previously indicated, allows core nations and corporations to distance themselves from many violent actions associated with natural resource extraction, actions that they might otherwise have to take themselves. This is critically important because when combined with citizens' lack of understanding of structural and sectoral adjustment, it allows these nations and corporations to shift blame for this violence (and the human rights abuses associated with it) away from themselves and to present their control over natural resources as the legitimate product of a just, rational, and efficient economic market. This, of course, further diverts core nation citizens' attention away from what core nation elites are doing and further shapes these citizens' understanding of the social and environmental implications of their nations' and leaders' policies and behavior.

The fact that extraction-related armed violence is often carried out by developing nations and rebels also helps to legitimate core nations, core nation corporations, international trade and finance institutions, and the global economic order by stigmatizing developing nations and rebels and disassociating core nations, core nation corporations, and the institutions they control from the violent underpinnings of the global extractive industry.[30] This disassociation, combined with the lack of readily available information on the social and environmental implications of World Bank–supported mining projects and reforms, also allows the World Bank to present itself as one of the world's foremost promoters of socially and environmentally progressive economic development (for instance, see Goldman 2005). Consistent with my fifth theoretical claim—that OINB inequality provides elites with the means to frame what is and is not considered to be pro-environmental behavior, policy, and development—the World Bank is thus able to use its organizational power to frame its actions and development policies as being good for the poor and the environment even as these actions and policies promote local and global environmental degradation, force hundreds of thousands of poor people to leave their homes, contribute little to the economic well-being of developing nations, and encourage the use of armed violence directed primarily against the poorest and most vulnerable people on the planet.

Finally, when one combines the evidence presented in this chapter with prior social science research on ecological unequal exchange and the direct environmental consequences of armed violence, militarism, and war, it quickly becomes apparent that armed violence and the environmental degradation and human rights abuses associated with it are intimately woven into the everyday lives of core nation citizens through the purchases they make and the fuels they consume.[31] It also becomes apparent that armed violence is a key driver of the global ecological crisis and that this is so because other key drivers of the crisis, such as the IMF, World Bank, WTO, and global marketplace, cannot, on their own, guarantee core nation access to and control over vital natural resources, the extraction of which is a key link in the chain connecting human activity and social organization to environmental degradation.

This, of course, suggests that armed violence is a serious environmental problem regardless of whether, in any specific instance, it is used, threatened, or merely implied. It also suggests that environmentalists and environmental social scientists have to examine and address armed violence and other resource extraction mechanisms simultaneously and in concert with other macro- and micro-level drivers of ecological degradation if they are to fully understand and solve the global ecological crisis. Finally, it means that institutions such as the World Bank, IMF, WTO, and global marketplace are violent both in their effects (unarmed violence) *and* because they require the use and threatened use of armed violence by other organizations, institutions, and networks in order to achieve their resource extraction and related capital accumulation goals. Thus, armed and unarmed violence are woven into the very fabric of the organizations, institutions, and networks that make elite power and elite goal attainment possible.

6

Restricted Decision Making and U.S. Energy and Military Policy in the George W. Bush Administration

On March 23, 2003, the United States and its allies commenced a "shock and awe" aerial bombardment campaign as part of their invasion of Iraq, in which coalition forces flew 15,500 strike sorties, dropped 27,000 bombs (Childs 2003), and in the first week of the war fired 675 Tomahawk cruise missiles (Associated Press 2003; Holguin 2009). This was not the first or last time the U.S. bombed Iraq or Iraqi forces. The U.S. military continued ground and air operations in Iraq until December 2011, and during the 1991 Gulf War the U.S. and its allies flew 40,000 strike sorties and dropped 265,000 bombs (Childs 2003). Moreover, between 1991 and 2003 the U.S., Great Britain, and France regularly bombed Iraq as part of their efforts to police two no-fly zones that they imposed on the country after the end of the first Gulf War (BBC News 2001; Turse 2011).[1] Indeed, between 1998 and 2000 bombing raids against Iraq occurred three to four times a week, possibly daily, with Britain dropping 780 tonnes (1,719,588 pounds) of bombs on Iraq and the U.S. flying 24,000 combat missions over southern Iraq during that period (Pilger 2000; Scahill 2000).[2]

These aerial operations, and the larger wars of which they were a part, devastated the people of Iraq, producing widespread death and destruction, severe social and economic dislocation, and government and infrastructural failure.[3] These wars also severely harmed the environment, both directly and by playing a key role in U.S. efforts to maintain access to and control over Iraqi and Middle Eastern oil, the burning and production of which produce tremendous quantities of greenhouse gases and other pollutants.

That these wars severely harmed the environment is not surprising given that war and preparation for war are among the most environmentally destructive of all human activities, contributing in important ways not only to the climate change crisis but also to a wide range of other environmental problems, including habitat destruction, biodiversity loss, soil degradation, and the generation and improper disposal of incredibly large volumes of hazardous, toxic, and radioactive waste (see the following section of the chapter).

What is surprising is that despite the key role that war and preparation for war play in harming the environment and despite the fact that the environmental movement views climate change as the most important problem facing the world, virtually none of the major environmental groups in the U.S. take an interest in the role that the U.S. military (the single largest user of energy in the world) and war play in causing the climate change crisis and otherwise harming the environment. Nor does the environmental movement seem aware of the strong linkages that U.S. leaders make between energy acquisition, military planning, and war. For instance, a quick perusal of the websites of several of the most prominent U.S. environmental organizations—the Environmental Defense Fund, the World Wildlife Fund, the National Wildlife Federation, the Sierra Club, the Natural Resources Defense Council, the Earth Policy Institute, Greenpeace, Friends of the Earth, and the Rainforest Action Network—shows that while they are all extremely concerned about climate change, only two of them (Greenpeace and the Earth Policy Institute) are concerned about the U.S. military and war, and even then it is only a minor concern in which war is framed as environmentally problematic solely because money spent on war and military preparedness cannot be spent on solving the world's most serious environmental problems.[4]

It is true that the Sierra Club in Canada posts a piece on its website about the environmental consequences of war that was written in 2003, that the Rainforest Action Network and Friends of the Earth are very concerned about human rights issues, and that the Natural Resources Defense Council and Greenpeace work to abolish nuclear weapons. It is also true that these environmental organizations all do a tremendous amount of advocacy and policy work on a wide range of critically important environmental issues. Nevertheless, none of these organizations appears to view military preparation, war, and war planning as integral to the environmental and climate change crisis despite the fact that the U.S. Department of Defense is probably the worst polluter in the world (Project Censored 2011) and that at least two of these environmental organizations—the Environmental Defense Fund and the Natural Resources Defense Council—praise the U.S. military profusely for its efforts to reduce its dependence on petroleum (Jordan 2012; Lehner 2011).

Given this lack of attention to military preparation, military planning, and war, one of the main goals of this chapter is to convince environmentalists that these factors are among the most important drivers of the global environmental crisis. However, understanding that these factors play a key role in harming the environment will not provide environmentalists with the knowledge and leverage they need to change U.S. policy if they are not also aware of how U.S. energy and military policy are formulated and linked to each other. The second goal of this chapter, then, is to examine how these policies were formulated and linked in the George W. Bush administration (2001–2009) and how they led the Bush administration to declare war on Iraq.

The third main goal of the chapter is to demonstrate that a small group of elite-controlled organizations, institutions, and networks played a key role in both shaping the Bush administration's energy and military policies and providing the administration with the means to hide from U.S. citizens the true nature of these policies and their links to each other.

As we shall see, the Bush administration was able to conceal the true nature of these policies from the public because its energy policy was created by an undemocratic White House task force, the National Energy Policy Development Group (NEPDG), and its military policy was formulated in a small group of neoconservative think tanks that had strong ties to the administration and to elite-funded neoconservative foundations. Consistent with my theoretical argument, the Bush administration thus relied on four undemocratic and elite-controlled mechanisms—the NEPDG, a small group of neoconservative policy planning organizations, the administration's links to these policy planning organizations, and overwhelming U.S. military superiority—to *formulate* and *achieve* its socially and environmentally damaging military and energy related policy goals.

Providing further support for my theoretical argument, these elite-controlled mechanisms provided the administration and neoconservative members of the power elite with the means to (a) monopolize decision making power regarding U.S. energy and military policy, (b) shift environmental and non-environmental costs onto others, (c) restrict the pro-environmental behavioral choices available to non-elites, and (d) divert the public's attention away from what the administration's true military and energy goals were and, hence, from what the administration was really doing in Iraq and around the world. These mechanisms were also used by elites to restrict public access to information that would have allowed citizens to democratically evaluate the administration's military and energy policies, thereby undermining citizens' democratic rights and decreasing the likelihood that they would oppose these policies in numbers large enough to obstruct elites' policy goals.

I turn now to a brief discussion of the human, social, and environmental consequences of heavy petroleum use and war.

The Human, Social, and Environmental Consequences of Heavy Petroleum Use and War

U.S. energy policy has committed the United States to heavy petroleum use since at least World War II (Klare 2004; Stoff 1980; Yergin 1991), with devastating human, social, and environmental consequences. One of the best known of these consequences is climate change, which among other things is expected to dramatically reduce the size and number of the earth's alpine glaciers, a key source of water for hundreds of millions of people, and melt significant volumes of ice in the arctic, the antarctic, and Greenland, raising sea levels anywhere from a few feet to several hundred yards depending on the severity of warming. Climate change is also expected to drastically alter ocean currents, which play an important role in shaping weather patterns, and depending on the region of the world in which you live, to increase or decrease temperatures and precipitation. It will also increase the likelihood of severe storms and prolonged drought (Cunningham and Cunningham 2007; Hawken et al. 1999; Speth 2005).

These changes, in turn, are likely to alter the global geography of agricultural production, increase infectious disease rates, and produce severe economic and social dislocation. They are also likely to force hundreds of millions of people to leave their homes as the ability to grow crops declines in certain parts of the world and as rising sea levels permanently flood the world's coastal areas, including many of the world's largest cities (including New York, London, Jakarta, and Manila) and much of Bangladesh and Pakistan, the world's sixth and seventh most populated countries (Cunningham and Cunningham 2007; Hawken et al. 1999; Speth 2005).

In addition to releasing vast quantities of greenhouse gases, petroleum exploration, extraction, processing, and combustion also produce a wide range of other waste products, including highly toxic drilling fluids, ground-level ozone, air particulate matter, thermal pollution, and wastewater contaminated with heavy metals, volatile aromatic hydrocarbons, and other toxic materials. Oil extraction and transport are also associated with oil spills and leaks that release tens of millions of gallons of petroleum into the environment each year (Epstein and Selber 2002, 4; Kharaka and Otton 2003; Klare 2004). For instance, over the past 50 years an estimated 540 million gallons of oil have been released into Nigeria's highly sensitive delta region (Nossiter 2010); between 1964 and 1990 Texaco deliberately released 345 million gallons of crude oil and 16–18 billion gallons of toxic waste into the Ecuadorian rainforest (Environment News Service 2012; Rainforest Action Network 2012; *Scandinavian Oil Gas Magazine* 2010); in 2010 the BP oil spill and explosion released 205.8 million gallons of oil into the Gulf of Mexico; and that same year the U.S. experienced "6,500 oil spills, leaks, fires, and explosions in addition to the BP oil spill and explosion. These spills, leaks, fires, and explosions released [an additional] 34 million gallons of crude oil and other potentially toxic chemicals into the environment, more than three times the amount released in the 1989 Exxon Valdez spill" (Keteyian 2011).

Petroleum extraction and transport are also strongly associated with military activity, armed violence, and war, both within nations and between them. Internal military conflicts over oil, including civil wars, separatist conflicts, and lower-level conflicts, have occurred, for example, in Colombia, Peru, Bolivia, Mexico, Ecuador, Nigeria, Angola, Iran, Russia, Indonesia, and Yemen, among other nations (Human Rights Watch 1999; Kovalik 2010: Ross 2010); and over the past 25 years the U.S. has fought major oil wars in Iraq and Kuwait and a war in Afghanistan that may have been driven, at least in part, by U.S. interest in building an oil pipeline through the country (Haslett 2001).

The U.S. also provides weapons, internal security hardware, and military and counterinsurgency training to its allies in oil-producing regions around the world, which are used by these allies to protect themselves and their or their region's oil from internal and external enemies (Klare 2004). Moreover, the U.S. military command structure, which consists of six *geographic* and three *functional* military commands, is designed with oil protection clearly in mind. Its African Command (Africom), for example, was created in 2007 to protect U.S. oil and mineral interests in Africa (Berschinski 2007), and its Central Command (Centcom) was created in 1983 to protect U.S. access to and control over oil in the Persian Gulf, the Caspian Sea basin, Southwest Asia, and until 2007 the Horn of Africa (Klare 2004). It is thus quite clear that the U.S. relies heavily on military power to ensure its continued access to and control over global oil supplies. It is also clear that many other countries use military force to protect oil production and transport within or near their own borders.

Given the critical role that military force and armed conflict play in ensuring the global flow of oil, it is impossible to separate the social and environmental consequences of global petroleum use from the social and environmental consequences of war, military preparation, and armed conflict, particularly in regard to the United States, which not only consumes more oil and spends more money on its military than any other nation in the world but which also led the oil wars in Iraq and Kuwait and the war in Afghanistan.

War, military preparation, and armed conflict are among the most environmentally destructive of all human activities (Downey et al. 2010; Project Censored 2011). The heavy machinery (trucks, tanks, etc.), bombs, and weaponry used during military training and combat operations destroy wildlife habitat, reduce biodiversity, destroy crops and arable land, leave highly toxic pollutants in the groundwater, soil, and vegetation, and severely degrade the soil by compacting and cratering it and producing extreme temperatures that kill important soil organisms (Clark 1992; Machlis and Hanson 2008). In addition, armies sometimes sabotage oil wells to achieve their aims and sometimes bomb chemical facilities and water treatment plants to weaken their enemy's morale and war-making capabilities, thereby releasing oil, chemicals, and human waste into the environment (Clark 1992).

For instance, near the end of the first Gulf War the Iraqi military intentionally detonated hundreds of Kuwaiti oil wells as it retreated from the country (Epstein and Selber 2002), though U.S. forces may have accidentally ignited some, perhaps many, of these wells (Clark 1992). These detonations resulted in the loss of an estimated 63 billion gallons of oil to fire, the spilling of an estimated 1–1.7 billion gallons of oil onto Kuwaiti soil (Epstein and Selber 2002), the estimated release of almost half a billion tons of carbon dioxide into the atmosphere (Carr 2007), and "the deposition of oil, soot, sulfur, and acid rain on croplands up to 1,200 miles in all directions from the oil fires, [making] fields untillable and [leading] to food shortages" (Jonathan Lash, quoted in Carr 2007, 338).

Like Iraq, the U.S. and its allies also destroyed important civilian and industrial infrastructure during the first Gulf War. For instance,

> Iraq's eight major multipurpose dams were repeatedly hit and heavily damaged [by U.S.-led forces]. This simultaneously wrecked flood control, municipal and industrial water storage, irrigation, and hydroelectric power. Four of Iraq's seven major water pumping stations were [also] destroyed. Bombs and missiles hit 31 municipal water and sewage facilities; 20 were hit in Baghdad alone. Sewage spilled into the Tigris and out into the streets of Baghdad . . . [and] in Basra, the sewage system completely collapsed. . . . [U.S.-led forces also bombed] 28 civilian hospitals, . . . 52 community health centers, . . . 676 schools, . . . three chlorine plants, a major ammonia export facility, 16 chemical, petrochemical, and phosphate plants, . . . 11 oil refineries, five oil pipeline and production facilities, and many oil tankers. (Clark 1992, 64–65, 66–67)

In addition, the U.S. fired 944,000 depleted uranium rounds during the war, which vaporized upon impact, releasing radioactive material that may have caused serious illness and birth defects among Iraqis, U.S. soldiers, and their children (C. Johnson 2004).

U.S. military installations also harm the environment.[5] J. R. McNeill and David Painter note, for example, that

> in addition to such special classes of lethal by-products as radioactive material, high explosives, chemical weapons, and rocket fuels, U.S. military bases generate[] massive amounts of hazardous waste . . . [as well as] used oil and solvents, polychlorinated biphenyls (PCBs), battery and other acids, paint sludge, heavy metals, asbestos, cyanide, and plating residues. Sometimes the size of small cities, U.S. bases also produce[]

large amounts of ordinary garbage, medical wastes, photographic chemicals, and sewage. . . . Training and maneuvers [at bases also] alter[] local
landscapes, consume[] vast amounts of energy, and contribute[] to air
pollution. (2009, 21–23)

It is not surprising, then, that in 2003 approximately 9,000 military
sites in the U.S. were contaminated (or believed to be contaminated)
with toxic and hazardous waste (EPA 2004) or that in 2010, 141 (8.7%) of
the United States' 1,620 National Priorities List Superfund sites, among
the most contaminated sites in the country, were owned by the Department of Defense (USGAO 2010), which several sources claim produced
more than 750,000 tons of hazardous waste per year in the early 2000s
(Project Censored 2004; St. Clair and Frank 2008).

U.S. military bases in foreign countries also produce tremendous volumes of waste. A Pentagon document leaked to the *Times* (UK) notes,
for example, that by 2009 U.S. troops had generated 11 million pounds
of hazardous waste in Iraq; and in 2010 a U.S. brigadier general stated
that he was "in the process of disposing of 14,500 tonnes [31.96 million
pounds] of oil and soil contaminated with oil . . . [that had] accumulated
[in Iraq] over seven years" (August 2010).

In addition, using data from 2009, the Army Environmental Policy
Institute (2010) estimated that Bagram Air Force Base in Afghanistan
generated 1.36 million pounds of hazardous waste and 677,644 pounds
of used oil per year, while the Institute of Medicine (2011) estimated
that during the wars in Iraq and Afghanistan, U.S. military bases with
more than 1,000 soldiers generated on average between 60,000 and
84,000 pounds of waste per day, with Joint Base Balad in Iraq generating between 200,000 and 400,000 pounds of waste per day. The Department of Defense provided the Institute of Medicine researchers with
little information regarding the composition of this waste, but the U.S.
Department of Veterans Affairs states that waste generated at bases in
Iraq and Afghanistan generally included "chemicals, paint, medical and
human waste, metal/aluminum cans, munitions and other unexploded
ordnance, petroleum and lubricant products, plastics and Styrofoam,
rubber, wood, and discarded food" (U.S. Department of Veterans Affairs 2012). Moreover, much of this waste was burned in open pits that
released toxic and hazardous contaminants directly into the air.

In addition to producing vast volumes of hazardous and nonhazardous waste, U.S. military operations in Iraq and Afghanistan, and U.S.-supported military operations in Pakistan, have destroyed forest cover, degraded marshlands and other wetlands, harmed local animal and bird populations, and contaminated water supplies with oil, depleted uranium, and chemicals such as benzene, trichloroethylene, and perchlorate. U.S.-led bombing campaigns and the intentional oil fires set by Saddam Hussein during the first Gulf War also polluted the air, soil, and water, while military operations and drought in Iraq, Afghanistan, and Kuwait produced dust laced with heavy metals such as arsenic, lead, cobalt, barium, and aluminum (Eisenhower Study Group 2011).

U.S. military operations in Iraq and Afghanistan also generated extremely large quantities of greenhouse gases, which is not surprising given that globally the U.S. military is the "world's single largest consumer of energy, using more energy in the course of its daily operations than any other private or public organization, as well as more than 100 nations" (Warner and Singer 2009, 1).[6] In 2006, for example, U.S. troops in Iraq and Afghanistan used roughly 57,000 barrels (2.39 million gallons) of oil *per day* (Crowley et al. 2007), well over half of it in Iraq. Indeed, energy use related to the war in Iraq was so great that in its first four years and nine months (from March 2003 to December 2007) the war generated 141 million metric tons of carbon dioxide equivalent, which at the time was comparable to the annual carbon dioxide emissions of 25 million U.S. cars and was greater each year than the annual emissions of most countries (during that time period there were 139 nations with lower carbon dioxide emissions than the war in Iraq; Reisch and Kretzmann 2008).[7] Moreover, this emissions estimate does not include greenhouse gases generated by the manufacture and detonation of explosives used in Iraq (Reisch and Kretzmann 2008), and as previously noted, it only covers the first four years and nine months of the war.

If we add the 141 million metric ton estimate highlighted in the preceding paragraph to the nearly 500 million tons of carbon dioxide estimated to have been released by oil well fires during the first Gulf War, and then add to this figure any greenhouse gas emissions from these two wars and the 12 years of enforcing the no-fly zones that were not included in these estimates; and if we then add to this all the hazardous and nonhazardous waste and other environmental harm generated by the United States' two decades of war with Iraq, including in our calculation all war-related emissions, waste, and degradation generated by weapons production, troop and material transport, and military training,[8] we cannot fail to realize what a tremendous environmental calamity these wars have been both in the region and globally. We also cannot fail to realize what a critical and central role war and militarism play in damaging the global environment, especially when we consider that the U.S. conducts military operations around the world, that all of the world's military forces harm the environment, and that weapons production and disposal, which I have not discussed, are also very environmentally damaging (Island Press 1999; Machlis and Hanson 2008).

However, the greatest cost of U.S. wars, and of war in general, is the toll they take on human lives. For instance, the use of depleted uranium munitions during the first Gulf War likely caused serious illnesses and birth defects among Iraqis, U.S. soldiers, and their children (C. Johnson 2004), and the burning of toxic and hazardous waste in open-air pits in Iraq and Afghanistan, the careless disposal of hazardous and toxic waste by U.S. soldiers in these countries, and the addition of heavy metals and dust to the air in Iraq, Afghanistan, and Kuwait likely increased the incidence of neurological disorders, respiratory problems, cardiovascular disease, and cancer among both soldiers and civilians (Eisenhower Study Group 2011; Kennedy 2008a, 2008b). Two decades of U.S. war and sanctions against Iraq also drastically increased unemployment and poverty in the country while decimating the country's public health and education systems and increasing youth illiteracy (Hassan 2005; IRIN 2006).

The wars in Iraq and Afghanistan also forced millions of people to flee their homes. By early 2007, for example, between 1.0 and 1.2 million Iraqis had fled to Syria as a result of the war (Syria placed this number at closer to 1.5 million), 750,000 had fled to Jordan, 150,000 to Lebanon, and 150,000 to Egypt, with an additional 1.9 million displaced Iraqis remaining in Iraq (Jamail 2007); and in 2012 there were still well over five million Afghani refugees living in Southwest Asia and around the world (UNHCR 2012), though presumably many of these refugees were displaced before the U.S. invasion of Afghanistan.

War also kills. In Afghanistan, 3,197 U.S. and coalition troops, 2,800 military contractors, and more than 10,000 Afghani soldiers and police officers had been killed as of October 2012, while 4,793 U.S. and coalition troops and 1,569 U.S. civilian contractors were killed in the war in Iraq (U.S. Department of Defense 2012; Foust 2012; iCasualties 2012; Tobin 2012). In addition, as of February 2013, official sources claimed that 50,476 U.S. soldiers and 25,839 U.S. contractors had been wounded in Iraq and Afghanistan, though it is likely that hundreds of thousands of additional U.S. soldiers experienced traumatic brain injuries and posttraumatic stress disorder as a result of these wars (Lutz 2013).

Civilian casualty figures are more difficult to estimate. Nevertheless, the United Nations reports that the war in Afghanistan resulted in 11,864 civilian deaths between 2007 and 2011 (United Nations Office of the High Commissioner for Human Rights 2012), while the organization Iraq Body Count (2012) estimates that between 109,564 and 119,701 Iraqi civilians died violently during the Iraq War.

Civilian death toll estimates such as these tend to be problematic, however, because they are generally based on incomplete data from sources such as morgue counts and media reports and because they generally fail to include people who die as a result of the social and economic devastation caused by war. In an attempt to overcome these shortcomings and provide a better estimate of the civilian death toll in Iraq, researchers from Johns Hopkins University and Al Mustansiriya University in Baghdad conducted a study that compared pre-war Iraqi death rates with Iraqi death rates between March 2003 (the beginning of the Iraq War) and July 2006. Using a random cluster sampling survey methodology that allowed them to interview families throughout much of Iraq, these researchers found that between March 2003 and July 2006 "there [were] 654,965 excess deaths [from all causes] in Iraq as a consequence of the war" (Burnham 2006).[9]

But even this figure likely underestimates the Iraqi death toll resulting from U.S. militarism because (a) some areas in Iraq where violence was especially high were difficult to survey, (b) the estimate does not include Iraqis who died after July 2006, and (c) in 2002, the baseline year for the study, Iraq had already suffered 12 years of war and sanctions, suggesting that the number of estimated "excess deaths" between 2003 and 2006 would have been higher if a more appropriate baseline year—1989 or 1990—had been used.[10] A complete record of Iraqi deaths resulting from U.S. militarism would also include all those who have died since 1990 as a result of the Iraq War, the Gulf War, and the sanctions and bombings imposed on Iraq between these two wars.

These, then, are among the true costs of war and militarism: death, dismemberment, loss of loved ones, the destruction of water and sanitation systems, social, educational, and economic devastation, psychological trauma, the creation of a refugee class often numbering in the millions, the burden placed on countries hosting war refugees,[11] and environmental devastation and destruction. The question that must therefore be asked is who decides when these costs are acceptable, and how and why do they decide to go to war? Or to phrase it more narrowly and manageably, given the role that the U.S. military plays in securing U.S. access to and control over global oil supplies and the environmentally and socially devastating consequences of the United States' war in Iraq, how and why did the Bush administration decide to go to war in that country, and how was this decision related to the administration's energy policy? It is to this question that I now turn.

The Bush Administration's National Energy Policy

The U.S. has been economically and militarily dependent on foreign oil since at least World War II, when it dramatically depleted its domestic oil supplies in order to win the war.[12] Indeed, President Roosevelt was so concerned about declining domestic oil reserves that he declared U.S. access to Saudi Arabian oil to be a vital national interest, and the centrality of oil to U.S. national interests and strategic planning has been reaffirmed by every U.S. president since then. Presidents Truman, Eisenhower, Nixon, and Carter, for example, all developed geopolitical doctrines designed either wholly or in part to protect U.S. oil interests in the Middle East, with each of these presidents pledging extensive military aid and training to U.S. allies in the region and Presidents Truman, Eisenhower, and Carter warning potential enemies that the U.S. was willing to use military force to protect its regional interests. President Carter even created a new military task force, the Rapid Deployment Joint Task Force (RDJTF), to enforce his doctrine, and in 1983 President Reagan expanded the RDJTF by creating a new geographic military command, Centcom, which was tasked with protecting U.S. oil interests in the Persian Gulf, the Caspian Sea basin, Southwest Asia, and the Horn of Africa (Klare 2004).

Like these presidents, President George W. Bush (2001–2009) viewed maintaining U.S. access to Middle Eastern oil as being in the national security interests of the United States. He and his advisers were also deeply concerned about the national security implications of two critically important oil-related issues: first, that without drastic increases in Middle Eastern and global oil production, oil producing nations would not be able to meet rapidly expanding global demand for oil, which would result in global oil shortages that could wreak havoc on the U.S. and world economies; and second, that due to 30 years of steadily declining oil production, the U.S. was becoming increasingly and dangerously reliant on a small number of oil-producing nations (Klare 2004).[13]

Thus, immediately after taking office, President Bush created the National Energy Policy Development Group (NEPDG) to develop a new energy policy for the United States. The NEPDG's report, the National Energy Policy (NEP), was released four months later, on May 17, 2001. While paying lip service to alternative energy sources and conservation, the report never reconsidered U.S. dependence on petroleum, which, in the context of drastically increasing global demand for oil, meant that global oil production would have to increase dramatically by 2025 in order for the U.S. and the world to remain economically stable (Klare 2004; NEPDG 2001). The overarching goals of the NEP were therefore to increase global oil production, "remov[e] . . . economic and political obstacles to overseas [oil] procurement," and diversify U.S. supplies of foreign oil (Klare 2004, 62). This would, of course, keep the U.S. dependent on foreign oil and the companies that sell it[14] but would hopefully improve national security by maintaining global economic stability and reducing U.S. dependence on any single foreign oil producer.

The NEP is notable not only for its insistence on maintaining U.S. petroleum dependence well into the 21st century but also for its relative silence regarding the degree to which its policy prescriptions would keep the U.S. dependent on *foreign* oil producers, its lack of candor regarding the economic, environmental, geopolitical, and military implications of continued oil dependence, and the secretive and undemocratic process used to prepare the report. I discuss the latter two of these issues in the sections that follow, starting with the undemocratic process the

NEPDG used to prepare the report and then moving to a discussion of the military implications of the NEP. Finally, I briefly examine the neo-conservative think tanks that helped develop the Bush administration's military policies, the neoconservative foundations that supported these think tanks, and the ties linking these think tanks and foundations to the Bush administration.

The NEPDG

The NEPDG was established on January 29, 2001 by order of the president and consisted of Vice President Dick Cheney, nine cabinet-level officials, and four senior administration officials, many of whom had ties to the oil and energy industries (see note 14). The NEPDG was charged with developing a plan that would ensure an adequate supply of energy for the nation in the future, taking into account potential supply disruptions, growing energy demand, environmental protection and conservation, and future energy infrastructure requirements (USGAO 2003).

In order for the NEPDG to gather the information it needed to accomplish its task, its members and staff met with hundreds of individuals from the private sector. However, it was very difficult for citizens and government officials outside the White House to determine exactly who the NEPDG met with and how influential these people were in shaping the NEP because the White House refused to disclose this information to the public. This lack of transparency led U.S. Representatives John Dingell and Henry Waxman to write a letter to the U.S. General Accounting Office (GAO) on April 19, 2001, requesting that the GAO investigate the NEPDG in order to determine, among other things, who the task force's members were, with whom they met, how much money the NEPDG was spending to develop its policy recommendations, and how it was developing these recommendations.

Unable to obtain the requested information, the GAO sued Vice President Cheney in U.S. district court in February 2002 to obtain the material it needed to fulfill Dingell and Waxman's request. The lawsuit was eventually dismissed on jurisdictional grounds, and Republicans threatened to cut GAO funding if the GAO pursued the lawsuit any further (Brand and Bolton 2003). This forced the GAO to base much of its investigation on limited information provided to it by the Environmental Protection Agency, the Office of Management and Budget, and the Departments of Energy, Interior, Transportation, Commerce, and Agriculture. However, even this information was released only under court order (USGAO 2003).

On the basis of this information, the GAO determined that the members and staff of the NEPDG met almost exclusively with high-level representatives of the petroleum, electricity, nuclear, coal, chemical, and natural gas industries (USGAO 2003). In addition, the *New York Times* discovered that although "more than 400 corporations and groups sought meetings with the energy task force, . . . [only] half that number were granted access, a group that included 158 energy companies and corporate trade associations, 22 labor unions, 13 environmental groups and a consumer organization" (Van Natta and Banerjee 2002, A19).

Moreover, the environmental groups that met with the task force did so only once, they were all present at the same meeting, and the meeting took place only after "the initial draft of the NEP was substantially complete" (Abramowitz and Mufson 2007). Thus, in developing the NEP, the NEPDG met almost entirely with energy industry officials, and when members of the U.S. Congress and the GAO attempted to determine whether this was true, how the NEPDG developed its policy recommendations, and how much money was spent to develop these recommendations, the White House prevented or made it very difficult for them to do so.

It should be fairly clear, then, that the Bush administration's decision making process was undemocratic in several respects, two of which were highlighted in this section and the third of which is the subject of the following section. First, the decision making process was nontransparent. As a result, U.S. citizens did not know how NEPDG decisions were made or who most influenced these decisions, making it difficult for citizens to determine whether these decisions were made democratically and who, if anyone, most benefited from them. Second, energy industry officials had much greater access to the NEPDG than did any other social or economic group in the United States. This not only undermined an important condition of democracy, that all citizens and interest groups have equal access to government decision makers (Pateman 1970), it also meant that energy industry officials were the only economic or social group in the nation that was able to provide significant input into the NEP's policy recommendations. As a result, the NEPDG can be viewed as an important elite-controlled mechanism that allowed energy industry executives to make their views known to the administration while preventing other groups from similarly influencing U.S. energy policy, thereby providing the administration and energy industry executives with the means to monopolize decision making power regarding U.S. energy policy.

Third, because the NEP failed to discuss the social, economic, environmental, geopolitical, and military implications of its recommendations, and because obtaining government documents describing these implications was difficult or impossible to do, U.S. citizens could not fully evaluate the NEP, making it easier for the administration to divert the public's attention away from its true policy goals and more difficult for citizens to determine whether they supported the Bush administration's energy policy. To illustrate this third point and discuss the roots of the Bush administration's *military* policy, the following three sections examine the military implications of the NEP and the role that neoconservative policy planning organizations played in the development of the Bush administration's military policy.

Military and Other Implications of the NEP

As previously noted, one of the most striking features of the NEP is that it failed to discuss several important problems inherent in trying to increase global oil production and diversify the United States' foreign oil supplies. One of the most critical of these problems was that if the U.S. was going to ensure an uninterrupted flow of oil not only to itself but also to the rest of the world, it would have to increase its military presence in virtually all the regions of the world that have significant oil reserves so as to ensure that enemies of the U.S. or of the governments of oil producing nations would not disrupt oil production and supply (Klare 2004). To do this, the U.S. would have to maintain or increase its already high levels of military spending, strengthen its economic and military ties to repressive governments in oil producing nations, and send its soldiers to fight in and/or train the armed forces of oil producing nations around the world (Klare 2004; Silverstein 2002).

Of course, money that the U.S. spends on the military cannot be spent on other important needs, such as reducing poverty, improving education, maintaining infrastructure, and developing clean alternative energy technologies; and increasing the United States' military presence in oil producing nations means ever-increasing foreign and American deaths, increased environmental destruction, and in some nations, social and economic chaos. Moreover, continued reliance on petroleum will worsen the climate change crisis and other environmental problems associated with fossil fuel use, extraction, and transport.

Another major problem associated with the Bush administration's energy policy was (and still is) that the vast bulk of the world's oil lies in the Middle East, which meant that for the Bush administration to achieve its goal of increasing global oil production, it would have to greatly increase Middle Eastern oil production (Klare 2004). This was problematic for several reasons. First, Iraq and Iran, with two of the largest oil reserves in the world, were both subject to U.S.-led economic sanctions when the Bush administration took office, making it difficult for international oil companies to invest in these key oil producing nations or for these nations to independently obtain the equipment they needed to maintain and increase oil production. Second, Iraq's military had the ability to disrupt oil production in the region, while Iran's military could (and still can) disrupt oil shipments out of the region. Third, French, Russian, Italian, and Chinese oil companies had signed contracts with Iraq to develop several of Iraq's oil fields once sanctions were lifted, so that removing sanctions on Iraq would not only leave Saddam Hussein in power, it would also put U.S. and British oil companies at a serious disadvantage in Iraq vis-à-vis their international competitors. Indeed, it is highly unlikely that U.S. and British companies would have ever received oil contracts in Iraq as long as Saddam Hussein was in power (Cafruny and Lehmann 2012; Klare 2004; *Economist* 2002).

The Bush administration's preferred solution to these problems, argues Michael Klare (2004), an expert on U.S. militarism and oil, was to replace the Iraqi and Iranian governments with regimes that were friendly to the U.S. This would open up both countries to foreign oil investment, make null and void any oil contracts signed by Saddam Hussein, and improve the competitive position of U.S. and British oil companies in Iran and Iraq, thereby aiding these oil companies and laying the groundwork for increased global oil production. It would also remove these countries as military threats to Middle Eastern oil production.

However, Iran was not internationally isolated and had an army that was likely to fiercely resist a U.S. invasion. As a result, the Bush administration decided that rather than invading Iran, it would increase pressure on the government while providing support to antigovernment activists in the country. Iraq, on the other hand, *was* internationally isolated with an army that was likely to surrender to U.S. forces relatively quickly. Klare (2004) therefore argues that the Bush administration viewed invading Iraq as both a viable option and a key strategic move in its attempt to achieve its energy related policy goals.

The Bush Administration's Military Policy

Despite the potentially serious consequences of the Bush administration's energy policy and the fact that policy alternatives did exist, the NEP failed to discuss these consequences or possible policy alternatives. The Bush administration was aware of many of these consequences, however. For instance, although the NEP did not discuss the military requirements of increasing global oil production and diversifying overseas oil procurement, the White House did direct the National Security Council (NSC) to "cooperate fully with the [NEPDG] as it considered the 'melding' of two seemingly unrelated areas of policy: 'the review of operational policies towards rogue states,' such as Iraq, and 'actions regarding the capture of new and existing oil and gas fields'" (Mayer 2004).

The NSC's assessment of the NEP's military implications is not publicly available. However, we do know that the Bush administration's military and national security policy called for a global Pax Americana, a U.S.-led global peace that was to be enforced militarily through the use of advanced weapons technology and enhanced power projection capabilities, capabilities that are crucial for protecting diverse sources of overseas oil (Barry and Lobe 2002; Klare 2004).[15]

Developed before the Bush administration came to power, this Pax Americana policy had its roots in neoconservative military thinking dating back to the early 1990s, when then-defense-secretary Dick Cheney directed Paul Wolfowitz to draft the 1992 Draft Planning Guidance (DPG), a set of classified guidelines written every two years that outline current U.S. military strategy (Mann 2004). The original draft of the 1992 DPG, which was greatly modified after it was leaked to the press, called for a world dominated militarily and politically by the U.S., a world in which no rival power could emerge in Europe, Asia, or the former Soviet Union to threaten our vital interests, including our interests in foreign oil (Tyler 1992). The draft DPG notes, for example, that the U.S.

> will retain the pre-eminent responsibility for addressing selectively those wrongs which threaten not only our interests, but those of our allies or friends. . . . Various types of U.S. interests may be involved in such instances: access to vital raw materials, *primarily Persian Gulf oil*; proliferation of weapons of mass destruction and ballistic missiles, [and] threats to U.S. citizens from terrorism or regional or local conflict. . . . In the Middle East and Southwest Asia, our overall objective is to remain the predominant outside power in the region and *preserve U.S. and Western access to the region's oil.* (*New York Times* 1992; emphases added)

The DPG was crafted by Zalmay Khalilzad, I. Lewis Libby, and Paul Wolfowitz, with important input provided by Richard Perle, Andrew Marshall, and Albert Wohlstetter (Mann 2004). Khalilzad, Wolfowitz, Perle, Libby, and Cheney were later associated with the Project for the New American Century (PNAC), which in 2000 published a report called *Rebuilding America's Defenses* (*RAD*). Building on many of the themes highlighted in the 1992 DPG, *RAD* likewise calls for protecting U.S. oil interests in the Middle East, bolstering U.S. power projection capabilities, and preserving a world in which the U.S. has no serious global or regional rivals:

> Today [the U.S. military's] task is to . . . deter the rise of a new great-power competitor; defend key regions of Europe, East Asia and the Middle East; and . . . *preserve American preeminence* . . . . America's global leadership, and its role as the guarantor of the current great-power peace, relies upon . . . the preservation of a favorable balance of power in Europe,

the Middle East and surrounding energy-producing regions, and East Asia. (Donnelly et al. 2000, 2, 5; emphasis added)

Many of the people associated with PNAC and its aggressive approach to military affairs entered the Bush administration when it came to power in 2001, including Dick Cheney, Donald Rumsfeld, Paul Wolfowitz, Zalmay Khalilzad, I. Lewis Libby, Richard Perle, Elliot Abrams, John Bolton, and Robert Zeollick; and five of these individuals—Cheney, Khalilzad, Wolfowitz, Perle, and Libby—were also involved in drafting the 1992 DPG. It should not be surprising, then, that the Bush administration strongly supported the aggressive military policies set forth in the PNAC and DPG reports or that it linked its military policies to the protection of Middle Eastern oil. The question still remains, however, of whether the Bush administration's military policy was tied in any way to the broader policy planning network highlighted by power structure researchers, in particular, to a group of neoconservative think tanks and foundations found within that network (see chapter 2 for a description of power structure theory).

Right-Wing Think Tanks and Foundations

Oil, advanced weaponry, enhanced power projection capabilities, and an aggressive military posture were clearly linked in the military strategies espoused by the Bush administration and the authors of the DPG and PNAC reports. Due to the overlapping think tank affiliations of key members of the Bush White House, Bush administration military strategy was also clearly (and directly) linked to the military strategies set forth in the DPG and PNAC reports and other reports produced by ultraconservative think tanks such as the American Enterprise Institute (AEI), the Center for Security Policy (CSP), and PNAC (Hartung and Ciarrocca 2003). Therefore, to fully understand the intellectual and organizational roots of the Bush administration's military strategy, we must briefly examine these ultraconservative think tanks.

Founded in 1943, AEI is one of the United States' most influential think tanks. It is also a highly conservative think tank whose stated purpose during the Bush administration was to "defend the principles and improve the institutions of American freedom and democratic capitalism—limited government, private enterprise, individual liberty and responsibility, [and] vigilant and effective defense and foreign policies" (AEI 2005). PNAC was founded in 1997 and closed its doors in 2005. Housed in the same building as AEI, PNAC described itself as being "dedicated to a few fundamental propositions: that American leadership is good both for America and for the world; and that such leadership requires military strength, diplomatic energy and commitment to moral principle" (PNAC 2008). CSP was founded in 1988. Its stated aim during the Bush administration was to ensure that the U.S. is "strong and prepared. We as a nation must . . . work to undermine the ideological foundations of totalitarianism and Islamist extremism with at least as much skill, discipline and tenacity as President Reagan employed against Communism" (CSP 2008).

During the Bush administration, PNAC, AEI, and CSP shared a number of common traits in addition to their strong conservative credentials. They all advocated achieving a Pax Americana through national missile defense, advanced weapons technology, and enhanced power projection; they were all funded by a common set of ultraconservative foundations; and they were all well represented in the Bush administration (Donnelly 2003; Dorrien 2004; Hartung and Ciarrocca 2003; Hossein-zadeh 2006; Lal 2005; MacDonald 2009; Muravchik 1991; Shah 2004; Vaisse 2010). For example, AEI and CSP each had at least 20 former employees, advisers, and close associates join the Bush administration in 2001 and 2002 (Bush 2003; Hartung and Ciarrocca 2003), and as previously noted, several prominent officials in the Bush administration were associated with PNAC in the late 1990s.

In addition, PNAC, AEI, and CSP all received a large portion of their funding from the Sarah Scaife, John M. Olin, and Lynde and Harry Bradley Foundations, with AEI also receiving significant funding from the Smith Richardson Foundation. These foundations, which were all created by wealthy businessmen and were subsequently controlled by these men, their families, or major corporations (People for the American Way 1996), channeled more than $107 million to conservative causes between 1999 and 2001, thereby providing ultraconservative members of the power elite with direct access to and influence over some of the most important ultraconservative policy planning organizations in the country (Krehely et al. 2004).

For example, in 2000, 64% of CSP's budget came from foundation grants (CSP 2001), and roughly 64% of this foundation money came from the Scaife, Olin, and Bradley Foundations. Similarly, about 44.5% of the $16.34 million that AEI received from foundations between 1999 and 2001 came from the Scaife, Olin, Bradley, and Richardson Foundations, and approximately 86.3% of the $1.43 million that PNAC received between 1997 and 2005 came from the Scaife, Olin, and Bradley Foundations (Foundation Center 2001, 2002; Media Transparency 2007).

AEI, PNAC, and CSP also received funding from individuals and corporations. For instance, in 2000 CSP received 9% of its revenue from individuals, 13% from defense industry corporations, and 14% from other corporations (CSP 2001); and in 1997 AEI received roughly one-quarter of its operating budget from foundations, one-quarter from corporations, one-quarter from individuals, and much of the remainder from conferences and sales (Toler 1999). Nevertheless, the importance of foundation money to the survival of these organizations was made clear in the 1980s when the Olin and Richardson Foundations cut off funding to AEI due to concerns that AEI was moving toward the center of the political spectrum. AEI soon changed its leadership and moved quickly to the right, only receiving renewed funding from these foundations after demonstrating its new ultraconservative credentials (Abelson 2006; Hershey 1986).

It thus appears that a very small group of ultraconservative foundations played a crucial role in guiding the basic policy directions taken by PNAC, AEI, and CSP in the late 1990s and the first few years of the 21st century. While it is highly unlikely that these foundations dictated specific policy positions to the think tanks they funded, it is clear that they were willing to stop funding think tanks that veered too close to the political center and that loss of funding pushed at least one think tank in a more conservative direction.

This small group of foundations also made a concerted and seemingly successful effort over the 30–40 years preceding the election of President George W. Bush to push U.S. politics in a very conservative direction (Krehely et al. 2004). They did this by (a) funding and developing a highly integrated network of ultraconservative organizations, (b) providing these organizations with general operating funds that could be allocated to whatever projects the recipient organizations deemed important, (c) providing funding for the long term so that these organizations did not have to worry about funding every year, and (d) providing funding for the entire range of activities necessary for political success (Covington 1997; Krehely et al. 2004).

As a result, AEI, CSP, and PNAC were all part of an ultraconservative policy planning network that was tied together and funded, and was to some degree created, by a small set of ultraconservative foundations that on the one hand were run by ultraconservative members of the power elite and on the other hand devoted considerable energy to promoting a highly conservative economic, political, and national security agenda. AEI, CSP, and PNAC were also tied to each other and to other conservative think tanks through overlapping organizational membership and their very strong ties to the Bush administration,[16] which adopted the military and national security policies that these think tanks and their affiliates had advocated since at least the early 1990s. Indeed, the Bush White House was, to a significant degree, made up of representatives and affiliates of these neoconservative think tanks, suggesting quite strongly that these think tanks were intimately involved in making U.S. military and energy policy in the Bush administration.

Inequality, Democracy, and the Bush Administration

The evidence presented in this chapter demonstrates that the Bush administration relied on four undemocratic and elite-controlled mechanisms—the NEPDG, a small group of neoconservative policy planning organizations, the administration's links to these policy planning organizations, and overwhelming U.S. military superiority—to formulate its military and energy related policy goals; make these goals U.S. policy; overcome economic, political, and military obstacles to expanded Middle Eastern oil production; and limit public access to information regarding the goals and consequences of its energy and military policies and how these policies were formulated and linked to each other. The evidence further demonstrates that a small number of ultraconservative foundations channeled key economic support from ultraconservative members of the power elite to think tanks whose paid affiliates not only helped develop the Bush administration's military policy but to a large degree became the Bush administration, thereby providing ultraconservative members of the power elite with an organizational and network-based mechanism for conveying their interests and policy goals to the White House.[17] The NEPDG likewise provided energy industry officials with the means to convey their interests and policy goals to the White House while excluding others from doing the same.

These organizational and network-based mechanisms thus provided the Bush administration, ultraconservative members of the power elite, and energy industry officials with the means to monopolize decision making power regarding U.S. energy and military policy and to restrict public access to information that would allow citizens to democratically evaluate U.S. energy and military policy. These mechanisms also provided the Bush administration with the means to invade Iraq, in part by allowing the administration to divert the public's attention away from the administration's true goals by making it difficult or impossible for the public to determine what these goals were—global military domination, continued petroleum dependence, and control over Middle Eastern oil—or what the social, military, and environmental consequences of the administration's energy policies were likely to be.

As a result, it was relatively easy for the Bush administration to garner citizens' support for its invasion of Iraq by falsely claiming that the goals of the invasion were to rid Iraq of weapons of mass destruction and punish Iraq for its ties to the September 11 terrorists, goals for which U.S. citizens were likely more willing to die (or send others to die) than the administration's true goals (the irony, of course, is that Iraq had neither weapons of mass destruction nor any ties to the September 11 terrorists).

Moreover, in making it difficult for the public to determine what role energy industry officials and ultraconservative members of the power elite played in developing U.S. military and energy policy, the NEPDG and the hidden organizational ties that existed between the Bush administration and ultraconservative foundations further diverted the public's attention away from what the administration was doing by making it difficult for citizens to determine who most influenced the Bush administration's policy and, therefore, who probably most benefited from it.

The Bush administration was thus able to shift the environmental and non-environmental costs of continued petroleum dependence onto U.S. citizens, Iraqi citizens, and all those who have been or will be negatively affected by U.S. oil wars, global climate change, and petroleum combustion, extraction, transport, and processing. The Bush administration was also able to limit U.S. citizens' pro-environmental behavioral choices by forcing young men and women to fight in an increasingly unpopular and environmentally destructive oil war and by directing government funding toward military preparedness and fossil fuel subsidies and away from alternative energy use and development (Bivens 2002).

The evidence presented in this chapter thus provides strong support for my argument that sociologists and environmentalists cannot fully explain the global environmental crisis without examining either the mechanisms that elites use to achieve their goals or the role that organizational, institutional, and network-based inequality (OINB inequality), undemocratic decision making, and elite-controlled organizations, institutions, and networks play in harming the environment. The evidence also strongly supports four of the six hypotheses I set forth in chapter 2, in particular, my predictions that OINB inequality is a fundamental source of environmental degradation because it (a) allows a small number of people and organizations to monopolize decision making power, (b) allows more powerful individuals, groups, and organizations to shift environmental costs onto less powerful individuals, groups, and organizations, (c) restricts the ability of non-elites to behave in environmentally sustainable ways by limiting the choices and shaping the incentives available to them, and (d) allows elites to divert the public's attention away from what they are doing so that their actions will not be scrutinized, questioned, or challenged or, in this case, so that the scrutiny, questions, and challenges are not so intense as to prevent elites from achieving their goals.[18] Finally, the general lack of awareness and concern among the public and major environmental groups regarding the extremely dire environmental consequences of war and militarism provides indirect support for another of my hypotheses, that OINB inequality inhibits the development and/or dissemination of environmental knowledge, attitudes, values, and beliefs, at least when this knowledge and these attitudes, values, and beliefs conflict with elites' perceived interests.[19]

Two possible objections might be raised regarding both my argument in this chapter and the ability of the IDE model to explain the events and problems described in it. First, it might be argued that U.S. militarism cannot be reduced to the Iraq War or to the protection of global oil supplies and that U.S. and global oil consumption cannot be explained solely by the decisions U.S. presidents make. I, of course, agree with this argument. Nevertheless, the decisions that presidents and government officials make about whether to provide government subsidies and support for fossil fuels, nuclear power, or alternative energy sources have important ramifications for the fuel mix actually used in this country, which is not insignificant given that the U.S. consumed roughly 25% of the world's oil when the U.S. invaded Iraq in 2003 (and roughly 20% today). The commitments that presidents make to continued petroleum dependence also have direct consequences for the military posture and geopolitical strategies the U.S. adopts as well as for the wars and military conflicts the U.S. starts and engages in. The U.S. military command structure is also designed with oil protection clearly in mind, and the Iraq War, which was primarily about oil, produced tremendous social, economic, and environmental devastation not only in Iraq but globally as well. It is difficult to claim, therefore, that the mechanisms and events described in this chapter did not (and do not) play an important role in shaping U.S. petroleum use, U.S. militarism, and the social and environmental problems associated with petroleum use, militarism, and war.

Second, it might be argued that rather than increasing the flow of oil from the Middle East and ensuring continued U.S. control over the region's oil, the Bush administration's invasion of Iraq weakened U.S. power in the region and did little to increase the flow of oil from Iraq. Until the summer of 2014 and the invasion of Iraq by the Islamic State of Iraq and Syria (ISIS), these were highly debatable points, with Iraqi oil production more than doubling between 2003 and 2013, most postwar oil contracts going to Anglo-American oil companies rather than to French, Russian, Italian, or Chinese oil companies, and most evidence pointing to continued U.S. military dominance in Iraq and the Middle East (Cafruny and Lehmann 2012; Kramer 2011). Moreover, only time will tell what the actual consequences of ISIS's invasion will be.[20]

But even if the U.S. invasion of Iraq did weaken U.S. power in the region, and even if ISIS's invasion undermines or further undermines U.S. power, dramatically decreases Iraqi oil production, and is the long-term result of Bush-era policies, this would have little bearing on my theoretical argument because I have never claimed that elites always correctly identify their interests or always pursue policies that are well designed (though in the case studies presented in this book they have done a very good job of doing so). Instead, I have argued that elites use undemocratic organizations, institutions, and networks that they control or dominate to achieve goals they have identified as being in their interests regardless of whether these goals and interests are actually achievable or good for them. It is thus clear that the evidence presented in this chapter strongly supports my theoretical argument.

It is also clear that environmentalists make a critical mistake in ignoring both the social and environmental devastation caused by militarism and war and the critical link that has existed since at least World War II between U.S. energy and military policy. Indeed, given the fact that military preparation, military planning, and war are among the most important drivers of the environmental and climate change crises, it is difficult to imagine how environmentalists will ever successfully address these crises without on the one hand severing the link that exists between U.S. energy and military policy and on the other hand reducing the United States' devotion to militarism and war. It is similarly difficult to imagine how they will overcome these crises if they do not drastically reduce OINB inequality and eradicate, weaken, or democratize the elite-controlled organizations, institutions, networks, and mechanisms highlighted in this chapter and throughout the book.

7

Environmental Degradation Reconsidered

Mainstream environmentalists offer a variety of solutions for the environmental crisis, including exhorting individuals and businesses to buy environmentally friendly goods, incorporating environmental costs into the prices of the goods businesses and consumers purchase, and incrementally developing a body of local, national, and international environmental law. Mainstream environmentalists also work actively to conserve critical species habitat, educate the public about important environmental issues, promote pro-environmental attitudes, values, beliefs, and behaviors, and develop new, environmentally friendly technologies and production processes.

With one exception, these are all potentially important approaches for solving the environmental crisis.[1] But as I pointed out in the introduction and chapter 1, they all share two critical shortcomings that put their ultimate success in doubt: they ignore the central role inequality plays in promoting environmental harm and they ignore the fact that social and environmental problems are usually caused by the same sets of factors and, thus, are intimately linked to one another.

Macro-structural environmental sociologists correct for these glaring errors in mainstream environmentalism by focusing their attention squarely on the links that exist between inequality, local, national, and global social structures, and negative social and environmental outcomes (see chapter 2 for a definition of macro-structural environmental sociology). But like mainstream environmentalists, macro-structural environmental sociologists pay insufficient attention to three important phenomena: organizational, institutional, and network-based inequality (OINB inequality), a very specific form of inequality that produces particularly severe social and environmental harm; war, militarism, and armed violence, which devastate people, communities, societies, and the environment around the world; and the inherently violent nature of elite-controlled organizations, institutions, and networks, which often rely on armed violence to achieve their goals and routinely pursue goals and outcomes that physically, emotionally, and psychologically harm people, communities, societies, and the environment.[2]

This book fills these important gaps in environmental thinking by demonstrating, across a wide variety of cases, that OINB inequality severely harms people, communities, societies, and environments around the world and that it does this by providing economic, political, military, and ideological elites with the power to develop and control a wide range of organizational, institutional, and network-based mechanisms that they use to accomplish six fundamental tasks. These tasks, which provide elites with the means to achieve goals that are often environmentally and socially destructive, and which are carried out using mechanisms that are often environmentally and socially destructive, include (a) monopolizing decision making power, (b) shifting environmental and non-environmental costs onto others, (c) inhibiting the development and/or dissemination of environmental knowledge, attitudes, values, and beliefs, (d) restricting the ability of non-elites to behave in environmentally sustainable ways, (e) framing what is and is not considered to be pro-environmental behavior, policy, and development, and (f) diverting the public's attention away from what elites are doing so that their actions will not be scrutinized, questioned, or challenged.

This book further demonstrates that each of the elite-controlled mechanisms it examines is associated with at least one of these six tasks, that each of these six tasks is associated with multiple elite-controlled mechanisms and with elite-controlled mechanisms operating in multiple empirical settings, and that elites regularly accomplish these six tasks as part of their efforts to achieve goals that are important to them, suggesting quite strongly that elites have to accomplish these tasks and use these mechanisms if they want to achieve their goals.

Moreover, because these mechanisms contribute in critically important ways to some of the most pervasive and damaging social and environmental problems the world currently experiences, including those associated with war, climate change, modern industrial agriculture, natural resource extraction, and IMF- and World Bank–led development, they play a key role not just in harming people, communities, and environments around the world, but also in producing the world's most serious social and environmental crises. It is thus clear that OINB inequality is one of the key drivers of these myriad crises and that this is so because it provides elites with the means to attain goals, such as accumulating capital, extracting natural resources, and dominating developing nations, that severely harm people, communities, and the environment and because the mechanisms that elites use to accomplish their goals are often environmentally, socially, and humanly destructive in and of themselves, as is the case with military power and armed violence.

As demonstrated in the preceding chapters, elites sometimes use a single mechanism to achieve their goals, but in most cases they either use multiple mechanisms to pressure non-elite actors to take a specific action, such as when the World Bank, IMF, and WTO simultaneously use free trade negotiations and structural adjustment loans to force developing nations to reduce their trade barriers, or they use overlapping mechanisms that are nested within each other and that operate and interact at multiple geographic levels to achieve their goals, as they did during the Latin American debt crisis, when economic and political elites in the U.S. used the U.S. policy planning network, power elite ties to the White House, structural adjustment loans, developing nation debt, and decision making rules at the World Bank and IMF to subordinate developing nations to their power.

Of course, if OINB inequality and overlapping, nested, and geographically dispersed elite-controlled mechanisms play a key role in harming people, communities, and the environment around the world, as I have demonstrated they do, then we cannot solve the global ecological crisis simply by incorporating social and environmental costs into prices, conserving habitat, developing new laws, treaties, and technology, or exhorting core nation consumers to change their purchasing behavior (as mainstream environmentalists would have us do). Instead, we must trace the flow of elite power back from the world's most serious social and environmental crises to the overlapping organizational, institutional, and network-based mechanisms that interact to create these crises. This will allow us to identify the organizational, institutional, and network-based mechanisms that must be altered or abolished if we are to solve the world's most serious social and environmental problems. It will also allow us to trace responsibility for these problems back to the national and global elites whose decisions ultimately created them.

However, in placing the lion's share of the blame for the world's myriad social and environmental crises on economic, political, military, and ideological elites, I do not mean to absolve middle class citizens and consumers from the obligation they have to help solve these crises. This obligation has three sources. First, it is quite clear, based on the evidence presented in this book, that elites are more interested in achieving their own goals than in protecting the environment or helping the world's poor, working, and middle classes. It is also quite clear that elites cannot be relied on to pursue goals that go against their financial interests or their interests in maintaining their organizational, institutional, and network-based power. Thus, if the world's social and environmental crises are to be solved, someone besides elites is going to have to solve them.

Second, middle class citizens in wealthy nations have a strong economic and moral connection to the people of the developing world. The armed violence, forced removal, abrogation of national sovereignty and democracy, destruction and poisoning of local environments, loss of wages and livelihoods, abuses of commodity chain power, enforcement of unfair trade rules, and poverty and immiseration that enable corporations operating in developing nations to pay low wages and inexpensively extract vast quantities of natural resources from the earth do not only enrich and empower economic, political, military, and ideological elites, they also make the goods and services that middle class citizens consume cheaper and more plentiful. So while it is true that middle class citizens have virtually no decision making power in the elite-controlled organizations, institutions, and networks highlighted in this book, it is also true that middle class citizens benefit, at least in some ways, from the decisions these organizations, institutions, and networks make. The middle class is, therefore, morally obligated to help those who are harmed by these decisions.

Third, many of the decisions that elite-controlled organizations, institutions, and networks make directly harm wealthy nation citizens. For instance, as I demonstrated in chapter 4, agribusiness firms use their power in agricultural commodity chains to reduce U.S. farmers' profits and farming options and produce unhealthy food. In addition, the lowering of trade barriers that I discussed in chapter 3 often pits U.S. workers against low-wage workers in other countries and U.S. businesses that operate domestically against companies that hire low-wage workers in other countries, resulting in lower wages, lost jobs, reduced profits, and failed businesses in the United States. The decision the Bush administration made to go to war in Iraq also harmed U.S. citizens, including the tens of thousands of U.S. soldiers and contractors who were killed and maimed in order to fulfill the administration's imperial ambitions, the hundreds of thousands of soldiers who returned home psychologically traumatized by their wartime experiences, and the civilians and soldiers who stayed home and spent anxious months or years worrying about their family and friends in Iraq.

The Iraq War and the war in Afghanistan also cost the U.S. $2 trillion, all of which was borrowed, with future expenses related to these wars, such as those associated with veterans' health care and disability benefits, expected to raise these wars' cost to between $4 trillion and $6 trillion (Bilmes 2013). This is debt that U.S. taxpayers will have to pay, and the money that is spent paying this debt is money that cannot be spent to rebuild our national infrastructure, improve our schools, create jobs, provide affordable health care, help the poor and unemployed, protect the environment, fund basic research, and develop alternative energy technologies. The existence of this debt, which was created by a supposedly fiscally conservative Republican president, will also likely be used by other "fiscally conservative" Republicans to justify cuts in government spending on important social and environmental programs.

U.S. citizens have also been harmed by the ties that exist between the power elite, the policy planning network, and the U.S. government. These ties, which I discussed throughout the book, have not only produced socially and environmentally damaging policies, they have also drastically undermined the democratic rights of U.S. citizens by giving the power elite and policy planning network disproportionate access to high-level U.S. decision makers and by providing the power elite and policy planning network with a mechanism that they alone possess for placing their members and employees in high-level positions in the U.S. government. As we have seen, this not only shuts most U.S. citizens out of the policy making process but, when combined with the secretive nature of much U.S. political decision making, makes it exceedingly difficult for U.S. citizens to determine which policies are in their and their country's best interests.

U.S. citizens are also directly harmed by the widespread environmental damage that elite-controlled organizations, institutions, and networks produce. While much of this damage occurs in the developing world, many of the environmental problems discussed in this book affect U.S. citizens, including climate change, soil erosion, the BP oil spill, the thousands of smaller oil spills, leaks, fires, and explosions that occur in the U.S. each year, and the rapid depletion of underground aquifers and loss of pollinator species that have occurred in the U.S. as a result of the widespread adoption of modern farming techniques. U.S. citizens are further harmed by the 500 million tons of manure that confined animal feeding operations in the U.S. produce each year, by the vast amounts of hazardous, toxic, and other waste that the U.S. Department of Defense and military contractors produce each year, by the drastic reduction in crop genetic diversity that has occurred since the rise of modern industrial agriculture, and by the more than one billion pounds of pesticides that are used in the country each year (Grube et al. 2011).

Elite-controlled organizations, institutions, and networks also play a key role in producing a host of social and environmental problems that I did not discuss in this book. For instance, ocean acidification, stratospheric ozone depletion, disruptions in the nitrogen and phosphorous cycles, changing land use patterns, biodiversity loss, atmospheric aerosol loading, and chemical pollution, which are produced by OINB inequality and a variety of other factors, all threaten the integrity of local and global ecosystems and therefore the interests of U.S. citizens, who depend on these ecosystems for their lives and sustenance (Foster et al. 2010). Organizational, institutional, and network-based mechanisms such as the World Trade Organization (WTO) and the North American Free Trade Agreement (NAFTA) also weaken the sovereignty of the U.S. and the democratic rights of U.S. citizens by allowing other countries (under WTO treaty provisions) and Canadian and Mexican companies (under NAFTA) to challenge the environmental, public health, and other laws and regulations of the United States. This results in many cases in fines or trade sanctions being imposed on the U.S., and in other cases, in the U.S. being forced to change its laws and regulations so as to satisfy the wishes of foreign countries and companies rather than the wishes of its own citizens (Public Citizen 2012, 2014; Wallach and Woodall 2004).

It is therefore clearly in the interests of wealthy nation citizens to ally with the rest of the world's citizenry in a global struggle against elite-controlled organizations, institutions, and networks and the socially and environmentally damaging decisions they make. The question that must be asked, then, is what must be done to weaken the power of elites, solve the world's myriad social and environmental crises, and improve the lives of the world's poor, working, and middle classes?

At the most abstract and obvious level, social and environmental activists need to drastically reduce OINB inequality within and between nations, drastically reduce the gap in decision making power that exists between elites and non-elites, and eradicate, weaken, or democratize the kinds of elite-controlled organizations, institutions, networks, and mechanisms highlighted in this book. But should this be done by overthrowing capitalism and the state, as some would advocate, or by changing these institutions from within?

The argument in favor of overthrowing capitalism and the state posits that capitalism and the state are inherently violent and antidemocratic and cannot exist without the exploitation of people, societies, and the natural world. Indeed, most of this book has been devoted to demonstrating that *as currently configured*, capitalism's dominant institutions are violent and antidemocratic and do routinely exploit people, societies, and the environment.[3]

It is difficult to imagine, however, that one can overthrow capitalism or the state without producing widespread violence and chaos and severe social and environmental devastation, thereby undermining the social and environmental goals of those who might be tempted to take such actions. And though some people would disagree, it is also difficult to imagine that large, complex societies can function without a state.

The solution to this dilemma may be for social and environmental activists to engage in a wide range of protest and other activities designed to democratize capitalism's currently dominant institutions as much as is possible, while simultaneously building and developing new economic, political, ideological, and cultural institutions that can eventually replace the institutions we currently have.[4] In other words, social and environmental activists should seek to increase non-elites' *control* of currently existing economic, political, ideological, and cultural institutions so as to (a) greatly improve social and environmental conditions around the world and (b) significantly weaken these institutions, with the goal, over time, of weakening them enough to eventually create a new, non-capitalist global political economy (and short of that to make the world a significantly better place), with the new political economy founded on democratic institutions that non-elites develop as they weaken the world's currently dominant institutions.[5]

Many people have written about the steps that must be taken to democratize our current national and global institutions and about the types of institutions that must be created to "transcend" the current capitalist political economy, including Heikki Patomaki, Teivo Teivainen, Erik Olin Wright, Archon Fung, Gene Sharp, Joshua Cohen, Joel Rogers, Bruce Ackerman, John Gastil, Peter Levine, Amy Lang, Boaventura de Sousa Santos, Robin Blackburn, Michael Albert, Marguerite Mendell, Phillipe Van Parijs, and John Curl.[6]

I will not duplicate their efforts here, but I will say that absolutely crucial to the success of a global democratization and empowerment project is the involvement of activists and citizens from all walks of life and all the nations of the world. Also crucial to the success of such a project is the reduction of OINB inequality not just within but also among nations (without such a reduction, powerful countries will continue to harm people, communities, and environments in less powerful countries by shifting the social and environmental costs of their actions onto these other countries). In addition, activists must make it increasingly difficult, and eventually impossible, for nations to benefit from war, militarism, and armed conflict, though reducing global inequality will, on its own, help significantly with such efforts by making it more difficult for nations to successfully use force against their own citizens and each other.

As is no doubt clear, my approach for solving the world's social and environmental crises is very different from the kinds of approaches taken by mainstream environmentalists and, thus, may not seem like environmentalism to many people. It will also be an extremely difficult approach to implement. Nevertheless, the argument and evidence presented in this book make it abundantly clear that OINB inequality is one of the key drivers of the world's myriad social and environmental crises and, as a result, that the only way to solve these crises is to drastically reduce OINB inequality and democratize or abolish the world's many elite-controlled organizations, institutions, and networks. Indeed, if we do not reduce OINB inequality and democratize or abolish these elite-controlled organizations, institutions, and networks, not only will powerful actors continue to shift the social and environmental costs of their actions onto others, they will also be able to prevent mainstream environmentalists from fully implementing the kinds of policies that they believe are required to solve the global environmental crisis. The approach I am advocating is, therefore, fundamentally environmental in nature.

It will, of course, be extremely difficult to force economic, political, military, and ideological elites to accept a more democratic, just, and equal world; and because these elites control the world's most important national and international decision making bodies, forcing them to do so will require social and environmental activists to adopt highly confrontational political tactics, such as mass demonstrations and other direct actions, that go outside normal political channels and place extreme pressure on elites and the organizations, institutions, and networks they control.

But as difficult as this will be, the only other alternatives are either to pursue less radical policies, such as those currently advocated by mainstream environmentalists, or to engage in even more drastic actions, such as overthrowing capitalism and the state. However, as I previously noted, it is impossible to imagine how one can overthrow capitalism or the state without producing widespread violence and chaos and severe social and environmental devastation; and on their own, the kinds of policies proposed by mainstream environmentalists have done little (and will do little) to solve the world's social and environmental crises because they are based on a glaringly faulty understanding of how the social, political, and economic world works and of how OINB inequality and elite-controlled organizations, institutions, and networks produce social and environmental harm. Thus, if we want to solve the world's myriad social and environmental crises, we have little choice but to drastically reduce OINB inequality and replace the world's elite-controlled organizations, institutions, and networks with democratic organizations, institutions, and networks that give voice to the voiceless and provide all the world's people with the power to protect themselves and the environment from the actions of others.

NOTES

INTRODUCTION

1. It is likely the case that many environmentalists who make this argument make it not because they fully agree with it but because they think it is the only approach to solving environmental problems that is politically viable. Nevertheless, since they act as if they believe this argument, their belief in it, whether merely professed or actual, has real social import.
2. For example, with the demise of colonial empires in the 20th century, powerful nations' control over developing nations and their natural resources could no longer be guaranteed through military might and colonial administration, but instead came to be ensured through a combination of developing nation debt, wealthy nation military power, and the actions of international institutions such as the World Bank, the International Monetary Fund, and the World Trade Organization, which collectively provide powerful nations with the means to strongly influence and sometimes dictate developing nations' domestic and foreign policies.

CHAPTER 1. POPULAR EXPLANATIONS OF THE ENVIRONMENTAL CRISIS

1. I do not disagree with the regulatory argument, but I do believe that its proponents generally fail to acknowledge just how unequal and undemocratic our political system really is and, as a result, just how difficult it will be to properly regulate industry. Moreover, proponents of this argument often promote a piecemeal approach to solving environmental problems that ignores larger social structures and power relations that must be drastically altered to make pro-environmental regulatory change (and other pro-environmental changes) more likely.
2. Many of the goods that businesses purchase are eventually sold to consumers, but many are consumed by these businesses themselves or altered and sold to governments or other businesses, which face a very different set of behavioral incentives than do individual consumers.
3. If companies do not recognize this, then they are not fully rational actors with perfect information.
4. My description of environmental economics is drawn primarily from Costanza et al. (1997), Endres and Radke (2012), Goodstein (2002), Hanley et al. (2013), Smith (2011), Thampapillai (2002), and Wiesmeth (2012).
5. Most environmental economists focus solely on environmental costs, but some incorporate social costs into their models too.

6. If a company is successfully sued for the social or environmental harm it causes, the costs of the lawsuit are likely incorporated into future prices or into future or past insurance premiums. However, these costs probably do not reflect the full social and environmental costs of the company's activities and are not necessarily distributed across the company's products in such a way as to reflect the actual social and environmental harm that each of the company's products causes.

7. There is even more to admire about environmental economics' more radical cousin, ecological economics. Ecological economics pays much closer attention than environmental economics to sustainability and justice and to the ethical, practical, and theoretical significance of the fact that the economy is a subsystem of the natural world, leading its practitioners to fundamentally rethink the role of the economy in society (Daly 1996). Environmental economists do recognize that the economy is a subsystem of nature, but unlike ecological economists, they do not acknowledge that this fact calls into question some of the most fundamental tools and assumptions of standard economics, tools and assumptions that also guide environmental economics. As a result, environmental economists attempt to solve the environmental crisis by tinkering with, rather than radically rethinking, economic policy and markets.

8. These bottlenecks "include trunk pipelines for gas, refineries for oil, railroad heads for coal, [terminals for] liquid natural gas, . . . [and] cement, steel, aluminum and [greenhouse gas]-intensive chemical plants" (Nader and Heaps 2008).

CHAPTER 2. INEQUALITY, DEMOCRACY, AND MACRO-STRUCTURAL ENVIRONMENTAL SOCIOLOGY

1. Portions of this chapter originally appeared in Downey and Strife (2010).

2. According to the theory, governments maintain their legitimacy by (a) maintaining a healthy economy that keeps people employed and (b) generating taxes that are used to provide government services to citizens and businesses.

3. Treadmill theorists have discussed international trade agreements and what they call "the transnationalization of the treadmill" in several places, and in one 20 year old article (Gould et al. 1995), they discuss international treaties in theoretical terms. However, in the most recent book-level treatment of treadmill theory written by the theory's main proponents (Gould et al. 2008), the authors do not incorporate international trade agreements into their theoretical model (talking about a subject is not the same thing as theorizing it). This is quite understandable, I would argue, since their theoretical interests lie elsewhere. But the fact still remains that international trade agreements do not play a central role in their theoretical model.

4. One exception to this is Robert Brulle (2000), who argues that a lack of democratic communication in modern capitalist societies prevents these societies from satisfactorily addressing the global ecological crisis. Another exception is Eric Bonds (2011).

5. "Non-environmental" world systems theory researchers have theorized several of these issues. However, unless I specifically note otherwise, all references I make to environmental WST refer solely to environmental WST and not to WST more broadly.

6. I recognize, for example, that Kenneth Gould et al. (2004, 2008) briefly discuss the role that the WTO and NAFTA play in fostering capital accumulation and environmental degradation and that Gould (1994) discusses the role that undemocratic government practices and state legitimation played in supporting treadmill processes in six communities in the U.S. and Canada. Similarly, I recognize that Andrew Jorgenson and Kennon Kuykendall (2008) discuss the important role that neoliberal institutions such as the IMF play in increasing foreign investment dependence and that Roberts and Parks (2007) emphasize the role that WTO side agreements play in shaping developing nations' climate-related bargaining positions. In other words, I recognize that treadmill scholars and ecologically oriented world systems researchers highlight elite-controlled mechanisms that are specifically related to their theoretical and empirical arguments. However, they do not theorize these mechanisms *as mechanisms* or highlight them as important theoretical constructs. Thus, as distinct conceptual categories, undemocratic institutions, elite-controlled organizational networks, and structures of accumulation do not play an important *theoretical* role in these works or in most treadmill or ecologically oriented world systems research.

7. Other theories, most importantly environmental WST, also allow researchers to use a single theoretical model to examine environmental degradation at multiple historical junctures.

8. Many of the world's dominant organizations and institutions claim to have adopted a pro-environmental ethic. However, claiming to have done so and actually doing so are very different things.

9. Four points: First, I am not arguing that elites form a completely unified social group. Instead, I believe that different groups of elites generally have different and competing goals, both within and across the four power networks. Second, because the absolute and relative power of these networks varies over time and across societies, which group of elites (e.g., economic versus political elites) has the most power within any society necessarily varies over time and from one society to another. Similarly, which group of national elites has the most international or global power also varies over time. Third, I do not mean to imply that non-elites form a single social group. Clearly, non-elites are divided into multiple social groups that often have competing interests and different levels of power due not only to race, gender, and class inequality but also to differential access to organizational and institutional power. Fourth, I recognize that elites and non-elites sometimes work together to achieve specific goals. Nevertheless, the amount of institutional, organizational, and network-based power possessed by elites is sufficiently large so as to place them in a structurally unique position that (a) gives them interests that generally differ significantly from those of non-elites and (b) allows them to achieve their interests much more often and more completely than is possible for non-elites.

10. Several theoretical approaches within environmental sociology link inequality, power, and weak democratic institutions to environmental degradation (for a brief discussion of these approaches, see Downey and Strife 2010, 155–156). Thus, what

makes the IDE model unique is not its focus on inequality, but rather its focus on elite-controlled mechanisms and the relationship between OINB inequality, elite-controlled organizations, institutions, and networks, and environmental degradation.

11. Although Boyce does not say it, this also includes the ability to shift costs onto the environment, which is largely powerless to prevent environmentally destructive behavior in the short run but not in the long run.

12. Two points: First, this does not mean that overcoming inequality will solve the environmental crisis all by itself. Second, the treadmill model, environmental WST, ecological unequal exchange theory, and other macro-structural environmental theories also hold that inequality and environmental degradation are inextricably linked to each other. However, Boyce makes this point somewhat more explicitly and in a more general sense than do these other theories.

13. These, of course, are not the only issues and questions that must be addressed if we are to fully understand and solve the world's myriad social and environmental crises.

14. Some people will argue that organizations, institutions, and networks that do not directly engage in armed violence are not violent. I clearly disagree with this argument, as do others (Gupta 2012). Nevertheless, the validity of the IDE model is not based on any particular definition of violence or on whether violence occurs in any specific situation.

15. It is my opinion that elites sometimes do intend to harm individuals, societies, and the environment but that very often they simply do not care if this is the result of their actions. I cannot prove this point, but it is difficult to imagine, for instance, that officials and organizations that have continuously imposed structural adjustment on developing nations since the early 1980s (see chapter 3) did not know either in the early 1980s or at some point in the following 30 years that structural adjustment devastates communities, societies, and the environment. By the same token, it is difficult to believe that U.S. political and military leaders are unaware of the human, social, and environmental consequences of fossil fuel use, torture, and intensive bombing campaigns (see chapter 6).

16. There are, of course, other methodological approaches I could have taken to evaluate the IDE model. For example, I could have designed the case studies so as to determine whether competing predictions drawn from the IDE model, environmental WST, the treadmill model, and ecological unequal exchange theory are supported or contradicted by the empirical evidence. Such an approach would have been problematic, however, because these theoretical models all focus on different aspects of the social structure. They also operate at different levels of abstraction, with the IDE model focusing on the mechanisms through which elites exert power and the other theoretical approaches focusing on more abstract phenomena such as long waves in the world economy, nations' structural positioning in the world system, and the structural correspondence between the interests of the state, labor, and capital. As a result, these theories' empirical predictions do not necessarily contradict each other, and testing them in conjunction with one another makes little sense.

CHAPTER 3. THE WORLD BANK, THE INTERNATIONAL MONETARY
FUND, AND THE ENVIRONMENT

1. The definition, existence, nature, and causes of globalization are widely debated in
   the literature. I avoid these debates in this chapter by focusing on the role that the
   World Bank and IMF play in promoting global economic integration rather than
   on the question of whether globalization is fact or fiction or whether it is synony-
   mous with global economic integration.

2. Two points: First, in defining neoliberalism as an organizational phenomenon, I am
   departing slightly from prior scholarship, which has focused considerable attention
   on the organizations through which neoliberal polices have been achieved without
   defining neoliberalism in organizational terms. Second, in arguing that the World
   Bank and IMF are neoliberal, I do not mean to imply that they have always been or
   always will be neoliberal.

3. As readers will no doubt notice, the arguments I make in this chapter bear some
   similarity to arguments made by scholars such as Sarah Babb (2009), Walden Bello
   et al. (1999), Philip McMichael (2008), and Ngaire Woods (2006), which should not
   be surprising given that we are all writing about the same substantive issues. However, unlike these (and many other) scholars who study the IMF and World Bank, I
   highlight the role that the U.S. power elite play in shaping U.S. policy regarding the
   World Bank, the IMF, and structural adjustment. Unlike these other scholars, I also
   use the evidence presented in this chapter to test my IDE theory and to highlight
   the critically important role that elite-controlled organizations, institutions, and
   networks play in harming people, communities, societies, and the environment.
   Thus, even when the specific arguments I make in this chapter are similar to those
   made by other researchers, my goal in presenting these arguments is different, as
   are the theoretical conclusions I draw from them.

4. The impetus for the Bretton Woods conference, and the design of the postwar
   economic order, came directly from the U.S. and the United Kingdom (UK), with
   the U.S. power elite playing a key role in shaping the U.S. position in its pre–Bretton
   Woods negotiations with the UK and the UK forced to accept much of what the
   U.S. wanted in these negotiations (Domhoff 1990).

5. The U.S. and UK initially wanted to create an International Trade Organization
   (ITO) that would form the third governance pillar of the new global economic
   order, but the U.S. Congress was unwilling to grant power to the ITO. As a result,
   global trade came to be governed by a much less powerful institution, the General
   Agreement on Tariffs and Trade (Peet 2003).

6. Countries experience balance of payments problems when the value of the goods,
   services, and financial capital that they import greatly exceeds the value of the
   goods, services, and financial capital that they export, adjusting for financial trans-
   fers to and from the country.

7. Fiscal policy refers to government spending and taxation policy; monetary policies
   are policies that affect a nation's money supply; the exchange rate refers to the value

of a nation's currency relative to other nations' currencies; and structural policies are policies that "remove barriers to productive investment and job creation. [Structural policies] cover a vast range of activities including strengthening banking systems, improving public sector and corporate governance, and reinforcing the rule of law" (Rieffel 2003, 49).

8. The IMF and World Bank would likely disagree with the claim that structural adjustment loans are imposed on developing nations, arguing instead that developing nations agree to these loans because these loans are designed to save these nations' economies. However, as I discuss later in the chapter, structural adjustment loans have not solved most adjusted nations' economic problems, most nations that accept these loans are in no position to go against the wishes of the IMF and World Bank, and the conditions that are included in these loans impinge greatly on the ability of national leaders to make important economic, social, and political decisions. It is thus highly unlikely that adjusted nations and their leaders freely enter into these loan agreements.

9. Two points: First, these studies control for a wide variety of factors that affect national economic outcomes. Second, in dictatorships, participation in IMF structural adjustment programs tends to result in increased government spending on health and education, though spending levels still remain low in comparison to democracies.

10. SAPRIN is the acronym for the Structural Adjustment Participatory Review International Network, an activist network fighting structural adjustment programs around the world.

11. Peet's book was released prior to the official publication of the SAPRIN study, but the text of the SAPRIN study was unofficially released in 2002, one year prior to the publication of Peet's book.

12. Another reason why mining sector reforms in developing nations often retard economic growth is that these reforms promote mining for export rather than for domestic processing and manufacture, activities that promote growth through the development of a more diversified and internally integrated national economy (UNCTAD 2005).

13. These factors distort these nations' political processes because officials in IMF- and World Bank–favored departments and agencies (a) have more resources than do their counterparts in other departments and agencies and (b) act as intermediaries between their nations and the IMF and World Bank, thereby serving as the key link between their nations and the two institutions that their nations rely on most for access to needed financial resources.

14. Although the World Bank publicly distanced itself from tropical forest and dam projects in the 1990s due to widespread public criticism of these projects, it recommitted itself to funding these types of projects in the early 2000s (Bosshard et al. 2003; Humphreys 2006).

15. The Carajas mine and railroad were also associated with several critical social problems, including armed violence, that I discuss in chapter 5.

16. The Tucurui dam was not funded by the World Bank, but it was central to the success of the Carajas Grande project. Therefore, in funding the port, mine, and railroad, the World Bank was implicitly giving its approval to the dam, even though it did warn that the dam would degrade the environment.

17. Researchers have only recently identified dam reservoirs as a potentially major contributor to climate change through their emissions of greenhouse gases. This new finding suggests that hydroelectric power is not necessarily a viable solution to the climate change crisis (World Commission on Dams 2000).

18. Activist organizations claim that dams funded by the World Bank have displaced more than ten million people. Though I have no reason to doubt this number, I have been unable to substantiate it.

19. These tonnage figures were derived from data used in Rich (2009).

20. Although the World Bank argues that the coal-fired plants it finances often produce electricity more cleanly than do more traditional fossil fuel-fired plants, the Environmental Defense Fund points out that larger emissions reductions could be achieved by financing renewable energy facilities.

21. One important thing the World Bank did to secure financing for the Nam Theun 2 dam was to provide $133 million in loan guarantees to other dam funders and investors (World Bank 2010). Loan guarantees, which are a form of insurance underwriting, play a key role in allowing the World Bank to leverage the financing needed to fund many of the development projects it promotes around the world.

22. The Nam Theun 2 increased river flow and flooding down the Xe Bang Fai River while decreasing water flow down the Nam Theun River (Lawrence 2008a).

23. This study is described in a book that Goldman (2005) wrote that examines a much broader range of World Bank–related topics than I summarize here.

24. The IMF is not similarly divided.

25. According to the World Bank, "If the country is a member of the Bank and is also a member of the International Finance Corporation (IFC) or the International Development Association (IDA), then the appointed Governor and his or her alternate serve ex-officio as the Governor and Alternate on the IFC and IDA Boards of Governors. They also serve as representatives of their country on the Administrative Council of the International Center for Settlement of Investment Disputes (ICSID) unless otherwise noted. Multilateral Investment Guarantee Agency (MIGA) Governors and Alternates are appointed separately" (World Bank 2012).

26. The executive directors of the IBRD, IDA, IFC, and MIGA are usually the same people (World Bank 2011a).

27. The size of these boards of directors has increased over time. For instance, the World Bank's board originally had 12 members, and on November 1, 2010, it added a 25th member. However, for the purposes of this chapter, which focuses primarily on events before 2010, I refer to the boards as having 24 members, which was the case from November 1, 1992, to November 1, 2010.

28. As Amy Horton notes, the lack of developing nation representation at the World Bank is particularly troubling given the fact that "developing countries represent

over 80 per cent of the world's population and [80% of the] Bank's membership, are where almost all of the Bank's activities take place; and, through loan repayments, are [one of] the main financial contributors to the Bank" (2010, 1). This lack of democratic accountability is, of course, equally troubling at the IMF.

29. The studies cited in this paragraph control for multiple factors that determine whether a nation receives favorable treatment from the World Bank and IMF.

30. The power that the U.S., UK, and western Europe hold at the Bank is also maintained by the Bank's English-only language policy, the Bank's culture of conformity, which includes conformity to a specific set of economic ideas, the lack of international diversity among the Bank's staff and management, and the homogeneous educational background of Bank staffers and management, most of whom have been trained at prestigious Western universities (Woods 2006; Yi-Chong and Weller 2009).

31. The World Bank and IMF attached conditions to their loans well before the 1980s (Kapur et al. 1997; Vreeland 2007), and the World Bank initially developed structural adjustment loans in 1979 and 1980 due to concerns internal to the Bank (Kapur et al. 1997). Nevertheless, as I demonstrate in this chapter, the World Bank and IMF would not have adopted structural adjustment as their key policy mechanisms for addressing the Latin American debt crisis if not for intense pressure by the U.S. (and to a lesser degree the UK and Germany).

32. The G-10 were the U.S., UK, Japan, Germany, France, Italy, Canada, Belgium, the Netherlands, and Sweden.

33. The Interim Committee, which became the International Monetary and Financial Committee (IMFC) in 1999, has the same number of members as the IMF board of directors and is selected in the same manner as the IMF board of directors. The committee reports to and advises the IMF's board of governors. However, as noted earlier, it is actually the G-7, rather than the IMFC, that guides the IMF (Woods 2006, 191).

34. Somewhat ironically, the World Bank was surprised by these developments, apparently having never been consulted by the U.S. about its new role prior to Baker announcing his plan and having been reminded by the U.S. only months before Baker's announcement that the U.S. planned on keeping the Bank on a short funding and lending leash for the foreseeable future (Kapur et al. 1997).

35. It is not clear which IMF executive directors raised objections to the plan on March 10, 1989, but given that Germany and the UK raised objections to it at the G-7 meeting, it is quite possible that it was these nations' representatives who raised objections on March 10.

36. It is important to note that since 1989 debt relief has never been carried out in any serious way, that it only results in partial debt forgiveness, and that it is almost always associated with the adoption of significant structural adjustment reforms by the nation receiving relief.

37. Unless otherwise noted, biographical information on these individuals is drawn from various editions of *Standard and Poor's Register of Corporations, Directors,*

and Executives, the 1985–1986 edition of *Who's Who in Finance and Industry*, and transcripts of these individuals' congressional nomination hearings.

38. I have been unable to determine when Brady joined the Council on Foreign Relations or when he stepped down from his positions at Rockefeller University and the Economic Club. So it is possible that he held these positions at the time of his 1988 nomination.

39. These differences probably did not seem small to moderate conservatives and ultraconservatives or to officials in the Reagan White House. But from the perspective of many people on the political left, these differences simply represented alternative ways of accomplishing the same basic goal, the domination of developing nations by the U.S., Europe, and Japan.

40. The argument set forth in *Mandate for Leadership* (Ture 1981, 685–686) and elsewhere for this dual policy was that the U.S. would be better able to influence developing nations if increases and decreases in foreign aid came directly from the U.S. rather than from proxy institutions such as the World Bank. Contradicting this position, others argue that funneling aid through the World Bank and IMF makes it easier for the U.S. to influence developing nations because the World Bank and IMF are viewed as being technocratic institutions without specific geopolitical interests. As a result, their policy prescriptions are viewed somewhat more favorably by developing nations than they would be if imposed directly by the U.S. (Woods 2006).

41. As a result, the story is not listed as a "negative environmental consequences" story.

42. Most of these stories cited the World Bank as the source of data used in the story.

43. The World Bank claims that since the 1990s it has provided affected people and communities with input into the projects that affect their lives; but as Goldman (2005) and SAPRIN (2004) demonstrate, this is not the case, and it certainly was not the case before the 1990s.

CHAPTER 4. MODERN AGRICULTURE AND THE ENVIRONMENT

1. Not everyone agrees with this assessment. Vandana Shiva (1991), for example, argues that agricultural productivity in the 20th century did not increase as dramatically as is widely held since comparisons of yields achieved on modern and "traditional" farms often ignore much of what is produced on traditional farms.

2. According to Bhoopendra Singh and Mudit Kumar Gupta (2009), pesticide exposure kills five million people per year.

3. A hectare is 10,000 square meters, which is approximately 2.47 acres.

4. The idea here is that because oligopoly and oligopsony firms have so many buyers or sellers to choose from, it is much easier for these firms to walk away from any potential market exchange than it is for their potential buyers or sellers to do so, giving these firms great power over their buyers and sellers.

5. As we shall see, oligopsony firms are also able to flood markets, though in this case to lower the prices of the goods that they buy from other actors.

6. Proprietary seeds are "branded [seed] varieties subject to intellectual property protections" (Hubbard 2009, 4).

7. Corn and soybeans are key ingredients in animal feed, which, in turn, is a key input for the livestock industry.

8. Price discovery refers to the determination of prices through the forces of supply and demand in the market. Price discovery cannot occur when the majority of transactions occur not in the marketplace, but through contracts signed by agribusiness firms and farmers.

9. A study funded by the U.S. Department of Agriculture found that "a 1 percent increase in the use of packer-ownership or contract production causes the spot market for hogs to fall by . . . 0.88 percent" (Hauter 2009, 25).

10. Integrating firms prefer to work with large growers because these firms need to run their processing plants at full capacity to remain cost competitive and because doing so reduces their transaction costs (Gurian-Sherman 2008).

11. These antibiotics are also used to promote animal growth.

12. As previously noted, proprietary seeds make up 82% of the global seed market (ETC Group 2008; Hubbard 2009, 4). The world's four largest seed companies thus controlled 43% of the global commercial seed market in 2007.

13. Cross-licensing agreements do not give the parties to the agreement ownership over the other company's property, but they do allow the companies to use each other's patented materials or techniques, usually without paying royalties for this use.

14. The one exception to this is genetically modified Bt corn, but only when European corn borer infestation is high (Gurian-Sherman 2009).

15. I define saving seed to plant new crops or develop new seed traits to be a production method that farmers using genetically modified seed can no longer use.

16. These governments are unable to subsidize agricultural production because free trade agreements do not allow them to do so (despite the fact that wealthy nations can offer such subsidies after signing these agreements), because the conditions attached to structural adjustment loans do not allow them to do so, or because they do not have the money to do so.

17. The AoA was also supposed to reduce domestic agricultural subsidies and agricultural export subsidies, but in the U.S. and Europe, domestic agricultural subsidies remain large (Murphy 2009). Moreover, the U.S. and Europe still spend significant sums on agricultural export subsidies, which (a) increase agricultural production, thus lowering the prices farmers receive for their goods, and (b) directly harm farmers who do not have access to these subsidies (Cairns Group 2012; USDA Foreign Agricultural Service 2015).

18. Amstutz claims to have played a major role in drafting the original negotiating text of the AoA while working at the U.S. Trade Representatives Office in the 1980s. He worked for Cargill both before and after his U.S. government service.

19. In the 1960s a very important and influential think tank, the Committee for Economic Development (CED), called for trade negotiations that would drastically reduce agricultural trade barriers (Hauter 2012). While it is likely that the work of the CED played a role in shaping the Williams Commission's views on agriculture, I do not have the evidence to prove this.

20. President Obama also supports agricultural trade liberalization (Kirk 2011).

21. Although many Europeans have struggled to keep genetically engineered food out of their markets, thus slowing down the spread of genetically modified crops, it is difficult to imagine that they will be able to successfully ban genetically modified food in the long run. Indeed, some European scientists are encouraging their nations to lift bans on genetically engineered food, and it is difficult to believe that bans on genetically engineered food will not at some point be successfully challenged at the WTO.

22. I am referring here to the organizational and network-based links that Edmund Pratt and the IPC used to mobilize U.S., European, and Japanese corporate support for an international property rights agreement. For a similar example of this, see Dreiling's (2001) book on NAFTA.

23. In this regard, it is important to recognize that (a) oligopsony and oligopoly firms are large organizations, (b) commodity chains are organizational networks, (c) liberalized agricultural trade occurs between actors in these organizational networks, (d) free trade agreements are institutional arrangements negotiated in and enforced by specific organizations, and (e) the U.S., the WTO, and developing nations are organizations embedded in overlapping national and international organizational networks.

CHAPTER 5. ARMED VIOLENCE, NATURAL RESOURCES, AND THE ENVIRONMENT

1. Portions of this chapter originally appeared in Downey et al. (2010).

2. Structural adjustment loans, for example, have produced or worsened poverty, hunger, disease, and social disorder—which can have devastating physical and psychological effects on individuals, families, and communities—in many countries around the world. Thus, structural adjustment loans and the elite-controlled mechanisms responsible for these loans (the World Bank, the IMF, the U.S. policy planning network, and the network's and power elite's ties to the White House) are violent in nature. The same is true for most, if not all, of the mechanisms highlighted in the preceding chapters, which even if they do not produce extensive harm on their own, often work in combination with other elite-controlled mechanisms to produce such harm.

3. I recognize that environmental sociologists who have written about militarism do not focus their attention solely on war or threatened war between nations. As stated in the main text, they also highlight the environmental consequences of weapons production, military activity, and war, and they argue that the relationship between these factors is important regardless of whether it is core, periphery, or semi-periphery nations that produce the weapons or engage in the military activity and war. However, when these scholars discuss the role that militarism plays in *ensuring core nations' access to natural resources*, what they generally highlight are differences in military power between core, periphery, and semi-periphery nations, indicating that what they have in mind is core nations' use of military threats, armed conflict, and war to gain and maintain access to developing nations' natural resources.

4. This argument shares several similarities with those put forth by Al Gedicks (2001), who discusses several of the issues I highlight in this chapter, and Michael Watts, Nancy Lee Peluso, and other political ecologists (Peluso and Watts 2001; Watts 2008). However, as important as Gedicks's work is, he does not situate his research within an overarching theoretical model or within a larger body of environmental sociology research; and political ecologists pay less attention than I do to the link between armed violence and core nation capital accumulation.

5. The TRI only tracks 650 chemicals, so this figure is probably low.

6. Colonialism is, of course, an important mechanism that powerful nations have used to exert control over natural resources (Foster 1994).

7. It should be clear from these examples (and from chapter 6) that I agree with the argument that core nation military power and military violence between nations both play an important role in ensuring core nations' access to natural resources in periphery nations. However, my argument is that these are just two of the ways in which armed violence contributes to environmental degradation and ecological unequal exchange.

8. Political risk insurance is insurance that companies can purchase to protect themselves against losses resulting from factors such as political violence (revolution, civil unrest, terrorism, war, etc.), government expropriation of assets, government repudiation of contracts, and government rules that limit their ability to repatriate profits and convert foreign currency.

9. Rebels and developing nation elites often benefit immensely from the sale of their nations' natural resources. Nevertheless, it is reasonable to think that they would behave differently, if not necessarily in their citizens' or the environment's best interests, if they had more power vis-à-vis corporations, core states, and international institutions such as the World Bank and IMF.

10. For instance, prior research shows that in Africa, foreign mining companies generally bring their own equipment and specialized employees to the continent, thereby ensuring that the sales for this equipment and much of the wages and skills for these jobs leave the continent. These companies also export the minerals they extract in African nations to other countries for processing and manufacture, pay exceedingly low royalty fees to African governments, due in large part to World Bank pressure on these governments, and often evade already low taxes (DanWatch and Concord Danmark 2010; UNCTAD 2005). Thus, it should come as no surprise that in mineral-rich African nations a sizeable increase in mining company profits between 2002 and 2006 was not associated with an increase in government mineral revenues (DanWatch and Concord Danmark 2010), that in Ghana in 2003 the government earned only 5.2% ($46.7 million) of the value of its total mineral exports ($893.6 million) (UNCTAD 2005), and that in Zambia, one of the poorest nations in the world, the government exempted Konkola Copper Mines from the nation's Mines and Minerals Act, which only requires companies to pay a 3% royalty rate. As a result, the company only paid a 0.6% royalty rate ($6.1 million) in 2006–2007 on more than $1 billion in revenue (Dymond 2007). The same basic trends have been identified in Latin America (Kumar 2009), especially in mineral-rich Peru,

where many multinational mining companies pay no royalties at all and where government revenues actually decreased in 2006–2007 despite skyrocketing mineral commodity prices (Salazar 2008).

11. I focus on minerals critical to the economy and national security of the United States because the U.S. has the largest economy and most powerful military in the world. Nevertheless, the minerals included in this study are critical to many core nation economies and militaries and to the global economy as well. For example, platinum, palladium, and rhodium are essential components in the production of automobiles around the world; manganese, niobium, vanadium, and titanium are necessary for global metal and alloy production; copper is an essential element in the global electronics and power production industries; and rare earth minerals are critical to global computer production (NRC 2008). Thus, the evidence presented in this section of the chapter has significance not only for the U.S. and the nations where the minerals are mined, but for other core nations and the global economy as well.

12. The year 2006 was the most recent year for which these data were available when I conducted this study.

13. The PTS provides two sets of scores, one based on information provided by Amnesty International and the other on information provided by the U.S. State Department. Since State Department evaluations may be shaped by U.S. geopolitical interests, I use the Amnesty International scores. In addition, because PTS scores can vary considerably from one year to the next due to fluctuations in the use of state violence, I use a 10-year average (1997–2007) to calculate each nation's PTS score. The one exception to this is Gabon, for which Amnesty International data do not exist for these years. Nevertheless, I define Gabon as having a highly repressive government on the basis of my reading of U.S. and NGO documents.

14. As should be clear from my prior discussion, I am not arguing that core nations do not use or support the use of armed violence to achieve their goals, that they do not or have not used armed violence in support of mining activities within their own borders, or that armed violence does not benefit core nations. In fact, an important element of my argument is that core nations and the institutions they control often shape the decisions that non-core actors make, including the decision to use armed violence, thus implicating core nations in these decisions. Moreover, to the extent that core nations (a) provide non-core nations or rebels with military equipment and training, (b) support repressive governments in non-core nations, or (c) directly take part in decisions regarding the use of armed violence in non-core nations, they are directly implicated in armed violence in these nations.

15. Companhia Vale do Rio Doce is now called Vale S.A.

16. Because the Azul manganese mine would not exist without the larger Carajas Grande development project, I consider any social or environmental problem associated with the Carajas Grande project to be associated with the Azul mine.

17. It is not clear whether electricity from the Tucurui dam is used by the Azul mine, but regardless of whether it is, the mine would not exist without the larger Carajas Grande development project, of which the dam is a critical component.

18. Other sources estimate that the mine produces 200,000–230,000 tons of waste a day.

19. Ivanhoe, which was (and may still be) partially owned by Rio Tinto, apparently sold its shares in the Monywa mine to members of Myanmar's ruling junta in 2009.

20. Myanmar's human rights violations at that time included

abridgement of the right to change the government; extrajudicial killings, including custodial deaths; disappearances; rape, torture, and beatings of prisoners and detainees; arbitrary arrest without appeal; politically motivated arrests and detentions; incommunicado detention; infringement on citizens' right to privacy; forcible relocation and confiscation of land and property; restriction of freedom of speech, press, assembly, association and movement; restriction of freedom of religion; discrimination and harassment against Muslims; restrictions on domestic human rights organizations and a failure to cooperate with international human rights organizations; violence and societal discrimination against women; forced recruitment of child soldiers; discrimination against religious and ethnic minorities; trafficking in persons, particularly of women and girls for the purpose of prostitution and as involuntary wives; restrictions on worker rights; [and] forced labor (including against children), chiefly in support of military garrisons and operations in ethnic minority regions. (U.S. Department of State 2006a)

21. Afghanistan has long been considered a potential site for a pipeline to transport oil to Western nations. It also apparently contains more than $1 billion in unmined minerals.

22. For example, in 2007 the U.S. provided Israel with $2.34 billion in military aid and Egypt with $1.3 billion in military aid; and in the two-year period from 2006 to 2007 the U.S. provided Saudi Arabia with $2.5 billion in military equipment transfers. The U.S. has also strengthened its military ties with and provided military aid to Nigeria, which is one of the United States' largest suppliers of oil (Akande 2011; Igbikiowubo 2004).

23. It is impossible to tell from the World Bank webpage, which was downloaded in 2011, when the report on structural adjustment spending since 1979 was published. Nevertheless, the report clearly covers World Bank loans from the 1980s and 1990s and possibly from the early 2000s as well.

24. This list of nations was derived from published accounts (Campbell 2004, 2010; Lambrechts 2009; McMahon 2010) and from a dataset I downloaded from the World Bank that lists developing nations that have received mining sector technical assistance loans. According to the World Bank, these loans are typically used to support mining sector reform (*Mining Journal* 2000). However, because the World Bank cloaks its activities in a fair bit of secrecy, it is impossible to definitively determine which nations should be included in this list and whether every nation included in the list should, in fact, be there.

25. Due to the capital intensive nature of the industry, Ghana's mining sector also lost more than 8,000 jobs between 1992 and 2000 (Campbell 2004).

26. Two points: First, to the best of our knowledge, the data we downloaded from the World Bank were complete. Second, we did not investigate a handful of IFC

loans that provided funding to mining services companies but not to specific mining projects.

27. We used several sources, including World Bank reports (McMahon 2010), NGO reports (Lambrechts 2009), and academic articles (Campbell 2004, 2010), to determine whether and when World Bank–supported mining sector reforms were first initiated in each of these countries, and we used the following list of mining sector reforms to help make these determinations: formulating a policy and strategy for the mining sector; reforming mining and taxation legislation; strengthening institutions and providing training of public- and private-sector officials; creating earth and environmental science database management systems; putting into place mining title registries and land management systems; improving the technical, environmental, and social conditions of small-scale miners; and privatizing state owned enterprises (*Mining Journal* 2000).

28. As noted earlier, ecological unequal exchange researchers have identified violence and threatened violence between nations as important mechanisms promoting ecological unequal exchange. However, armed violence is a much broader concept than this, encompassing, but going well beyond, military violence between nations.

29. Research demonstrates that the mainstream U.S. news media is owned and controlled by a small handful of corporations and elites who are tied to non-news-media corporations and elites in a variety of ways (Noam 2009; R. Rice 2008).

30. The importance that this type of "disassociation" plays in legitimizing core nation and corporate activities is highlighted by the difficulties the diamond industry faced when the violence associated with diamond mining became highly publicized. It was also highlighted by the Bush administration's repeated attempts to convince U.S. citizens that its invasion of Iraq was not about oil.

31. This is true regardless of whether core nation citizens are aware of or in any way responsible for this violence.

CHAPTER 6. RESTRICTED DECISION MAKING AND U.S. ENERGY AND MILITARY POLICY IN THE GEORGE W. BUSH ADMINISTRATION

1. France stopped patrolling the no-fly zones in 1998.

2. Most of these U.S. combat missions took place over populated areas, such that "in one five-month period, 41 per cent of casualties were civilians: farmers, fishermen, shepherds, their children and their sheep" (Pilger 2000).

3. Although a strong case can be made that U.S. military efforts against Iraq really constituted a single, two-decade war, for the purposes of clarity I treat the two-decade war as though it was two wars with a long, low-level conflict in between.

4. Two points: First, I looked at these websites in the summer and fall of 2012. Second, Greenpeace is clearly concerned about the morality of war, but it seems to highlight this concern primarily before and during major U.S. military endeavors rather than on a regular basis.

5. In 2011 the U.S. operated 135 large military installations around the world, defined as installations for which the cost of replacing facilities and supporting infrastruc-

ture was at least $1.715 billion. Of these large installations, 111 were located in the U.S., 21 in foreign countries, and three in U.S. territories. One of the largest of these installations, Ramstein Air Force Base in Germany, occupied 3,102 acres in 2011 and had an estimated replacement value of $3.74 billion, while Kadena Air Force Base in Japan occupied 4,914 acres and had an estimated replacement value of $5.8 billion. In 2011 the U.S. also operated 122 medium-sized and 4,444 small installations around the world, defined, respectively, as installations with an estimated replacement value of between $915 million and $1.715 billion and between $0 and $915 million; 18 of the medium-sized and 571 of the small installations were located in foreign countries.

6. In 2006 the U.S. Department of Defense consumed more electricity than Syria and only slightly less than Denmark, placing its electricity use ahead of all but 57 nations (Lengyel 2007); and in 2009 it used approximately 395,000 barrels of oil per day, or 6.05 billion gallons per year, roughly equivalent to all the oil consumed in Greece that year (Warner and Singer 2009).

7. Nikki Reisch and Steve Kretzmann's (2008) calculations are rough estimates based on (a) information they had available to them and (b) carbon dioxide equivalent conversion factors associated with the combustion of different types of fuel. In addition to fuel used by troops, their calculations include greenhouse gases emitted as a result of troop deployment flights, cement production related to wartime activities, gas flaring above pre-war levels, oil burned in oil well fires, and upstream refining activities.

8. Reisch and Kretzmann (2008) include emissions from troop deployment flights in their estimate for the four years and nine months of the Iraq War they examined.

9. U.S. forces did not kill all these people. But the U.S. started the war and is therefore responsible for creating the situation within which these people died.

10. Such an estimate would have broadened the scope of the study to include the effects of all U.S. military activity and sanctions against Iraq since 1990 on excess deaths between March 2003 and July 2006.

11. As of 2007 the U.S. had done virtually nothing to help those Middle Eastern countries that were hosting the more than two million Iraqi refugees then living within their borders; and as of April 24, 2007, the U.S. had granted only 466 visas to Iraqi refugees (Jamail 2007).

12. Much of the argument in this and the next three subsections is borrowed from Michael Klare (2004), though my discussion of how this argument relates to democracy and to my theoretical model is entirely my own.

13. Under the Obama administration, U.S. oil production has increased to such a degree that the share of U.S. oil consumption derived from domestic oil supplies is higher than at any point in recent memory, with the U.S. potentially poised to become a net oil exporter by 2030 (International Energy Agency 2012). However, oil is a finite resource, so U.S. oil production will decline again at some point in the not too distant future. There is also reason to believe that projections regarding future U.S. oil production may be overly optimistic (Kobb 2013). But most importantly for the

argument presented in this chapter, when the Bush administration decided to invade Iraq, U.S. oil production had been decreasing and its energy consumption increasing continuously since the 1970s, with global oil demand also increasing rapidly.

14. In this regard, it is important to note that a number of prominent people in the Bush administration had economic ties to the oil industry, including President Bush (he and his father both ran independent oil companies), Vice President Dick Cheney (former CEO of Halliburton), Commerce Secretary Donald Evans (former CEO of Tom Brown Inc. and holder of oil industry stock), National Security Advisor Condoleezza Rice (former board member of Chevron), Defense Secretary Donald Rumsfeld (consultant for PB-Amoco), and EPA administrator Christine Todd Whitman (she had holdings in oil companies and oil producing properties) (Kay 2001; *Multinational Monitor* 2001). During the 2000 presidential campaign the Bush administration also received $2.8 million from the energy industry (Puzzanghera 2001), $1.9 million of which came from oil interests (Palmer 2002).

15. In this context, power projection refers to the ability of U.S. military forces to rapidly and effectively respond to military and national security threats anywhere in the world.

16. Officials in the Bush administration were also affiliated with the Jewish Institute for National Security Affairs, the RAND Corporation, the Hudson Institute, the Hoover Institution, the Committee for the Liberation of Iraq, Freedom House, and Empower America, among other policy planning groups.

17. Members of the power elite also sat on the boards of trustees of these ultraconservative think tanks, in particular, on the board of AEI.

18. It should be noted that many U.S. citizens accurately understood much of what the Bush administration was trying to achieve. However, the extreme secrecy with which the administration operated, the general lack of public knowledge regarding how U.S. energy and military policy were formulated and linked in the strategic planning of neoconservative think tanks and administration officials, and the willingness of the Bush administration to lie or distort the truth about Iraqi weapons of mass destruction and Iraq's ties to terrorists meant that those citizens who understood what the Bush administration was trying to accomplish often did not have the evidence they needed to substantiate their claims and convince others that they were right. In addition, their arguments were not given nearly the level of coverage and respect in the mainstream news media as those of their opponents, making it even more difficult for them to present their case to the general public.

19. Although I have demonstrated that mainstream environmental organizations in the U.S. largely ignore the environmental consequences of war and militarism, I have not demonstrated that this is the case for the U.S. public. However, given how extremely difficult it was for my research assistants and me to find research and evidence regarding the environmental consequences of militarism and war, it seems quite reasonable to conclude that U.S. citizens know very little about these consequences.

20. I am writing this in April 2015 and so do not know what the long-term ramifications of ISIS's invasion will be.

CHAPTER 7. ENVIRONMENTAL DEGRADATION RECONSIDERED

1. In chapter 1, I argued that environmental economics and exhorting people and businesses to purchase environmentally friendly goods are both fundamentally flawed because they do not account for the role inequality plays in shaping social and environmental outcomes. Based on other arguments I made in chapter 1, I find it difficult to believe that exhorting individuals to change their purchasing behavior can ever have much of an environmental impact. However, under the right conditions, many of the policies proposed by environmental economists could potentially produce positive environmental outcomes.

2. Please recall from the introductory chapter and chapter 2 that macro-structural environmental sociologists discuss many important elite-controlled organizations, institutions, and networks in their research. However, as theoretical constructs, OINB inequality and elite-controlled organizations, institutions, and networks play almost no role in macro-structural environmental sociology. Please also recall (from chapter 5) that relatively few environmental sociologists highlight the role that war, militarism, and armed violence play in harming the environment.

3. I would like to thank David Pellow for pointing out that the elite-controlled organizations, institutions, and networks highlighted in this book are not just undemocratic, they are actually antidemocratic.

4. This approach to solving the world's social and environmental crises is borrowed in very general terms from Erik Olin Wright (2010).

5. Non-elites will also have to exert control over the world's police and military forces, but without using these forces to dominate and control other people.

6. Most of these authors are cited in Wright's book *Envisioning Real Utopias*. Wright does a wonderful job of clearly summarizing their various arguments.

REFERENCES

ABC News. 2012. "BP Oil Spill: Two Years Later, Dispersants' Effects Still a Mystery." July 10. http://www.firstcoastnews.com/news/article/263639/4/BP-oil-spill-Two-years-later-dispersants-effects-still-a-mystery.

Abelson, Donald. 2006. *A Capitol Idea: Think Tanks and US Foreign Policy.* Montreal: McGill-Queen's University Press.

Abramowitz, Michael, and Steven Mufson. 2007. "Papers Detail Industry's Role in Cheney's Energy Report." *Washington Post,* July 18, A1.

Advertising Age. 2010. "Global Marketers 2009." November 30. http://adage.com/datacenter/globalmarketers09.

AEI (American Enterprise Institute). 1977. "Food and Agricultural Policy." Washington, DC: AEI for Public Policy Research.

———. 2005. "AEI's Organization and Purposes." http://web.archive.org/web/20090212000018/http://www.aei.org/about/ (accessed April 13, 2015).

Africa News. 2007. "Guinea: Strike Spells More Hardship." January 25. LexisNexis Academic.

Africa News Service. 2003. "MMCZ, Murowa in Diamonds Agreement." *Financial Gazette,* December 19. LexisNexis Academic.

Agence France-Presse. 2011. "Vietnam Says Laos Suspends Mekong Dam Project." International Rivers, May 9. http://www.internationalrivers.org/resources/vietnam-says-laos-suspends-mekong-dam-project-2730.

Agencia de Informacao de Mocambique. 2010. "Families Transferred from the Moatize Coal Mining Site." May 4. LexisNexis Academic.

Akabzaa, Thomas 2004. "African Mining Codes a Race to the Bottom." *African Agenda* 7 (3). http://www.choike.org/nuevo_eng/informes/2400.html.

Akande, Laolu. 2011. "US Strengthens Nigeria's Security, Donates Two Warships." *Guardian,* May 13. http://odili.net/news/source/2011/may/13/15.html.

Alba, Joseph, and J. Wesley Hutchinson. 2000. "Knowledge Calibration: What Consumers Know and What They Think They Know." *Journal of Consumer Research* 27:123–156.

Allard, Tom. 2005. "Indonesia Accused of Papua Atrocities." *Sydney Morning Herald,* August 19. LexisNexis Academic.

Amnesty International. 1998. "Corumbiara and Eldorado de Carajas: Rural Violence, Political Brutality and Impunity." January 1. http://www.amnesty.org/en/library/asset/AMR19/001/1998/en/dom-AMR190011998en.html.

Andersen, Thomas, Henrik Hansen, and Thomas Markussen. 2006. "US Politics and World Bank IDA-Lending." *Journal of Development Studies* 42:772–794.

Army Environmental Policy Institute. 2010. "Fully Burdened Cost of Managing Waste in Contingency Operations: Final Technical Report." Arlington, VA: Army Environmental Policy Institute.

Ascher, William. 1992. "The World Bank and US Control." In *The United States and Multilateral Institutions: Patterns of Changing Instrumentality and Influence*, edited by Margaret P. Karns and Karen A. Mingst, 115–140. London: Routledge.

Associated Press. 2003. "6,000 Precision-Guided Bombs Dropped on Iraq since War Began." Fox News, March 29. http://www.foxnews.com/story/0,2933,82559,00.html.

August, Oliver. 2010. "America Leaves Iraq a Toxic Legacy of Dumped Hazardous Materials." *Times* (UK), June 14. http://www.greenchange.org/article.php?id=5927.

Babb, Sarah. 2009. *Behind the Development Banks: Washington Politics, World Poverty, and the Wealth of Nations.* Chicago: University of Chicago Press.

Barkema, Alan, David Henneberry, and Mark Drabenstott. 1989. "Agriculture and the GATT: A Time for Change." *Economic Review* 74:21–42.

Barry, Ellen. 2014. "After Farmers Commit Suicide, Debts Fall on Families in India." *New York Times*, February 22. http://www.nytimes.com/2014/02/23/world/asia/after-farmers-commit-suicide-debts-fall-on-families-in-india.html?_r=0.

Barry, Tom, and Jim Lobe. 2002. "U.S. Foreign Policy—Attention, Right Face, Forward March." Silver City, NM: Foreign Policy in Focus.

Bartley, Tim. 2007. "How Foundations Shape Social Movements: The Construction of an Organizational Field and the Rise of Forest Certification." *Social Problems* 54:229–255.

Battikha, Anne-Marie. 2002. "Structural Adjustment and the Environment: Impacts of the World Bank and IMF Conditional Loans on Developing Countries." Blacksburg: Urban Affairs and Planning, Virginia Polytechnic Institute.

BBC News. 2001. "No-Fly Zones: The Legal Position." February 19. http://news.bbc.co.uk/2/hi/middle_east/1175950.stm.

———. 2008. "'Deaths' in Guinea Mine Protest." October 10. http://news.bbc.co.uk/2/hi/africa/7663573.stm.

———. 2013. "Mozambique Protesters at Brazil-Owned Vale Coal Mine." April 17. http://www.bbc.co.uk/news/world-africa-22191680.

Behere, P. B., and M. C. Bhise. 2009. "Farmers' Suicide: Across Culture." *Indian Journal of Psychiatry* 51:242–243.

Bello, Walden F., Shea Cunningham, and Bill Rau. 1999. *Dark Victory: The United States and Global Poverty.* Oakland, CA: Pluto.

Benbrook, Charles. 2009. "Impacts of Genetically Engineered Crops on Pesticide Use in the United States: The First Thirteen Years." Boulder, CO: Organic Center.

Berschinski, Robert 2007. "Africom's Dilemma: The Global War on Terrorism, Capacity Building, Humanitarianism, and the Future of U.S. Security Policy in Africa." Carlisle, PA: Strategic Studies Institute, U.S. Army War College.

Biello, David. 2010. "The BP Spill's Growing Toll on the Sea Life of the Gulf." Yale Environment 360, June 9. http://e360.yale.edu/content/feature.msp?id=2284.

Billig, Jennifer, and David Wallinga. 2012. "United States Farm Bill 2012: What's at Stake?" Minneapolis, MN: Institute for Agriculture and Trade Policy.

Bilmes, Linda. 2013. "The Financial Legacy of Iraq and Afghanistan: How Wartime Spending Decisions Will Constrain Future National Security Budgets." Cambridge: Harvard Kennedy School.

Binet, Les, and Peter Field. 2009. "Empirical Generalizations about Advertising Campaign Success." *Journal of Advertising Research* 49:130–133.

Bird, Graham. 2001. "IMF Programmes: Is There a Conditionality Laffer Curve?" *World Economics* 2:29–49.

Bird, Graham, and Dane Rowlands. 2001. "IMF Lending: How Is It Affected by Economic, Politician and Institutional Factors?" *Journal of Policy Reform* 4:243–270.

Bivens, Matt. 2002. "Fighting for America's Energy Independence." *Nation*, April 15.

Blanco, Humberto, and Rattan Lal. 2010. *Principles of Soil Conservation and Management.* New York: Springer.

Bloomfield, Steve. 2006. "Dutch Court Jails Liberian Warlord's Right-Hand Man." *Independent*, June 8. LexisNexis Academic.

Bonds, Eric. 2011. "The Knowledge-Shaping Process: Elite Mobilization and Environmental Policy." *Critical Sociology* 37:429–446.

Bosshard, Peter, Janneke Bruil, Korinna Horta, Shannon Lawrence, and Carol Welch. 2003. "Gambling with People's Lives—What the World Bank's New 'High-Risk/High-Reward' Strategy Means for the Poor and the Environment." New York: Environmental Defense Fund, Friends of the Earth, and International Rivers Network.

Boughton, James. 2001. *Silent Revolution: The International Monetary Fund: 1979–1989.* Washington, DC: International Monetary Fund.

Boyce, James K. 2002. *The Political Economy of the Environment.* Northampton, MA: Edward Elgar.

Brand, Peter, and Alexander Bolton. 2003. "GOP Threats Halted GAO Cheney Suit." *The Hill* (blog), February 19. http://thehill.com/news/021903/cheney.aspx.

Branford, Sue. 2007. "Fanning the Flames: The Role of British Mining Companies in Conflict and the Violation of Human Rights." London: War on Want.

Braun-LaTour, Kathryn, Michael LaTour, Jacqueline Pickrell, and Elizabeth Loftus. 2004. "How and When Advertising Can Influence Memory for Consumer Experience." *Journal of Advertising* 33:7–25.

Brookings Institution. 1973. "Toward the Integration of World Agriculture." Washington, DC: Brookings Institution.

Brown, Jonathan. 2006. "Niger Delta Bears Brunt after 50 Years of Oil Spills. *Independent*, October 26. http://www.independent.co.uk/news/world/africa/niger-delta-bears-brunt-after-50-years-of-oil-spills-421634.html.

Brown, M. 2010. "Letter to the New York Times Editor." *New York Times*, June 12. LexisNexis.

Bruckner, Monica. 2011. "The Gulf of Mexico Dead Zone." Northfield, MN: Microbial Life Educational Resources, Carleton College. http://serc.carleton.edu/microbelife/topics/deadzone/index.html.

Brulle, Robert J. 2000. *Agency, Democracy, and Nature: The U.S. Environmental Movement from a Critical Theory Perspective*. Cambridge: MIT Press.

Buira, Ariel. 2003. "The Governance of the IMF in a Global Economy." In *Challenges to the World Bank and IMF: Developing Nation Perspectives*, edited by Ariel Buira, 13–36. London: Wimbledon.

Bunker, Stephen. 1984. "Modes of Extraction, Unequal Exchange, and the Progressive Underdevelopment of an Extreme Periphery: The Brazilian Amazon, 1600–1980." *American Journal of Sociology* 89:1017–1064.

———. 2005. "How Ecological Uneven Development Put the Spin on the Treadmill of Production." *Organization & Environment* 18:38–54.

Bunker, Stephen, and Paul Ciccantell. 2005. *Globalization and the Race for Resources*. Baltimore: Johns Hopkins University Press.

Burbach, Roger, and Patricia Flynn. 1980. *Agribusiness in the Americas*. New York: Monthly Review Press.

Burnham, Gilbert, Shannon Doocy, Elizabeth Dzeng, Riyadh Lafta, and Les Roberts. 2006. "The Human Cost of the War in Iraq: A Mortality Study, 2002–2006." Baltimore and Baghdad: Bloomberg School of Public Health at Johns Hopkins University and the School of Medicine at Al Mustansiriya University.

Burris, Val. 1992. "Elite Policy Planning Networks in the United States." In *Research in Politics and Society: The Political Consequences of Social Networks*, vol. 4, edited by Gwen Moore and J. Allen Whitt, 111–134. Greenwich, CT: JAI.

———. 2005. "Interlocking Directorates and Political Cohesion among Corporate Elites." *American Journal of Sociology* 11:249–283.

Bush, George W. 2003. "Speech on the Future of Iraq." Hilton, Washington, DC.

Cafruny, Alan, and Timothy Lehmann. 2012. "Over the Horizon? The United States and Iraq." *New Left Review* 73:5–16.

Cairns Group. 2012. "Export Subsidies: Detrimental to Developing Country Exports." http://cairnsgroup.org/DocumentLibrary/export_subsidies.pdf (accessed July 23, 2012).

Campbell, Bonnie, ed. 2004. "Regulating Mining in Africa: For Whose Benefit?" Uppsala, Sweden: Nordiska Afrikainstitutet.

———. 2010. "Revisiting the Reform Process of African Mining Regimes." *Canadian Journal of Development Studies* 30:197–217.

Canfield, Sabrina. 2012. "Doctor Says Symptoms of Gulf War Illness and Oil Spill Illness Are 'Identical.'" Courthouse News Service, March 20. http://www.courthouse-news.com/2012/03/20/44836.htm.

Carlock, Catherine 2010. "Ivanhoe Breaks Contract with Rio Tinto." *Market Watch*, July 13. http://www.marketwatch.com/story/ivanhoe-stock-jumps-on-break-with-rio-tinto-2010-07-13.

Carlson, Ann. 2009. "Offsets and Waxman Markey." *LegalPlanet* (blog), July 31. http://legalplanet.wordpress.com/2009/07/31/offsets-and-waxman-markey/.

Carolan, Michael S. 2008. "Making Patents and Intellectual Property Work: The Asymmetrical 'Harmonization' of TRIPS." *Organization & Environment* 21:295–310.

———. 2011. *The Real Cost of Cheap Food*. New York: Earthscan.

Carr, Paul. 2007. "'Shock and Awe' and the Environment." *Peace Review: A Journal of Social Justice* 19:335–342.

Carstensen, Peter C. 2004. "Buyer Power and Merger Analysis—The Need for Different Metrics." Hearings held on February 17 by the U.S. Antitrust Division and Federal Trade Commission, Washington, DC.

Center for Food Safety. 2005. "Monsanto vs. U.S. Farmers." Washington, DC: Center for Food Safety.

———. 2007. "Monsanto vs. U.S. Farmers: November 2007 Update." Washington, DC: Center for Food Safety.

Chachage, C. S. L., Magnus Ericsson, and Peter Gibbon. 1993. "Mining and Structural Adjustment: Studies on Zimbabwe and Tanzania." Uppsala, Sweden: Nordiska Afrikainstitutet.

Chang, Ha-Joon. 2008. *Bad Samaritans: The Myth of Free Trade and the Secret History of Capitalism*. New York: Bloomsbury.

Childs, Nick. 2003. "America's Air War on Iraq." BBC News, April 15. http://news.bbc.co.uk/2/hi/2950837.stm.

Choi, Byung-Chul, Hang-Sik Shin, Su-Yol Lee, and Tak Hur. 2006. "Life Cycle Assessment of a Personal Computer and Its Effective Recycling Rate." *International Journal of Life Cycle Assessment* 11:122–128.

Chowla, Peter. 2007. "Double Majority Decision Making at the IMF: Implementing Effective Board Voting Reform." London: Bretton Woods Project.

Clark, Brett, and John Bellamy Foster. 2009. "Ecological Imperialism and the Global Metabolic Rift." *International Journal of Comparative Sociology* 50:311–334.

Clark, Brett, and Andrew Jorgenson. 2012. "The Treadmill of Destruction and the Environmental Impacts of Militaries." *Sociology Compass* 6:557–569.

Clark, Ramsey. 1992. *The Fire This Time: U.S. War Crimes in the Gulf*. New York: Thunder's Mouth.

Cohn, Martin. 2001. "Canadian Firms Justify Close Ties to Generals." *Toronto Star*, February 2. LexisNexis.

Committee on Banking, Housing, and Urban Affairs, U.S. Senate. 1979. *Hearing before the Committee on Banking, Housing, and Urban Affairs, United States Senate, Ninety-Sixth Congress, First Session, on the Nomination of Paul A. Volcker to Be Chairman, Board of Governors, Federal Reserve System*. Washington, DC: U.S. Government Printing Office.

Cornford, Jonathan. 2008. "A Greater Mekong Subregion? Reflecting on 16 Years of the ADB's GMS Initiative." *Watershed* 12:44–50.

Cornford, Jonathan, and Nathaniel Matthews. 2007. "Hidden Costs: The Underside of Economic Transformation in the Greater Mekong Subregion." Victoria, Australia: Oxfam Australia.

Costanza, Robert, John Cumberland, Herman Daly, Robert Goodland, and Richard Norgaard. 1997. *An Introduction to Ecological Economics*. Boca Raton, FL: St. Lucie.

Covington, Sally. 1997. "Moving a Public Policy Agenda: The Strategic Philanthropy of Conservative Foundations." Washington, DC: National Committee for Responsive Philanthropy.

Cox, Robert. 1979. "Ideologies and the New International Economic Order: Reflections on Some Recent Literature." *International Organization* 33:257–302.

Croome, John. 1999. *Reshaping the World Trading System: A History of the Uruguay Round.* Boston: Kluwer Law International.

Crowley, Thomas D., Tanya D. Corrie, David B. Diamond, Stuart D. Funk, Wilhelm A. Hansen, Andrea D. Stenhoff, and Daniel C. Swift. 2007. "Transforming the Way DoD Looks at Energy: An Approach to Establishing an Energy Strategy." Mclean, VA: LMI Government Consulting.

CSP (Center for Security Policy). 2001. "2001 Annual Report." Washington, DC: CSP.

———. 2008. Home page. http://www.centerforsecuritypolicy.org.

Cunningham, William, and Mary Cunningham. 2007. *Environmental Science: A Global Concern.* New York: McGraw-Hill.

Curran, Sarah. 2012. "Domestic Production vs. Imports for Rice in Haiti: A Delicate Balance to Strike." Center for Sustainable Development. http://cgsd.columbia. edu/2012/06/16/domestic-production-vs-imports-for-rice-in-haiti-a-delicate-balance-to-strike/ (accessed July 24, 2012).

Dabbs, W. Corbett. 1996. "Oil Production and Environmental Damage." Washington, DC: TED. http://www1.american.edu/ted/projects/tedcross/xoilpr15.htm.

Daly, Herman. 1996. *Beyond Growth: The Economics of Sustainable Development.* Boston: Beacon.

DanWatch and Concord Danmark. 2010. "Golden Profits on Ghana's Expense: An Example of Incoherence in EU Policy." Copenhagen: DanWatch and Concord Danmark.

Davenport, Christian. 2007. *State Repression and the Domestic Democratic Peace.* New York: Cambridge University Press.

David, Laurie. 2008. *Stop Global Warming: The Solution Is You!* Golden, CO: Fulcrum.

Democracy Now. 2012. "Gulf Oil Spill: BP Execs Escape Punishment as Fallout from Disaster Continues to Impact Sea Life." April 23. http://www.democracynow. org/2012/4/23/gulf_oil_spill_bp_execs_escape.

Dertouzos, James N., and Steven Garber. 2006. "Effectiveness of Advertising in Different Media." *Journal of Advertising* 35:111–122.

De Schutter, Olivier. 2009. "Seed Policies and the Right to Food: Enhancing Agrobiodiversity and Encouraging Innovation." New York: United Nations.

Desmarais, Annette. 2003. "The WTO . . . Will Meet Somewhere, Sometime. And We Will Be There!" Ottawa, ON: North-South Institute.

Dolan, Catherine, and John Humphrey. 2001. "Governance and Trade in Fresh Vegetables: The Impact of UK Supermarkets on the African Horticulture Industry." *Journal of Development Studies* 37:147–176.

———. 2004. "Changing Governance Patterns in the Trade in Fresh Vegetables between African and the United Kingdom." *Environment and Planning A* 36:491–509.

Domhoff, G. William. 1990. *The Power Elite and the State: How Policy Is Made in America*. New York: Aldine de Gruyter.

———. 2002. *Who Rules America? Power and Politics*. New York: McGraw-Hill.

Domina, David A., and C. Robert Taylor. 2010. "The Debilitating Effects of Concentration Markets Affecting Agriculture." *Drake Journal of Agricultural Law* 15:61–108.

Donnelly, Thomas. 2003. "What's Next? Preserving American Primacy, Institutionalizing Unipolarity." Washington, DC: American Enterprise Institute, May 1. http://www.aei.org/publication/whats-next-2/.

Donnelly, Thomas, Donald Kagan, and Gary Schmitt. 2000. "Rebuilding America's Defenses: Strategy, Forces and Resources for a New Century." Washington, DC: Project for the New American Century.

Dorrien, Gary. 2004. *Imperial Designs: Neoconservatism and the New Pax Americana*. New York: Routledge.

Downey, Liam, Eric Bonds, and Katherine Clark. 2010. "Natural Resource Extraction, Armed Violence, and Environmental Degradation." *Organization & Environment* 23:417–455.

Downey, Liam, and Susan Strife. 2010. "Inequality, Democracy, and the Environment." *Organization & Environment* 23:155–188.

Drahos, Peter. 2002. "Developing Countries and International Intellectual Property Standard-Setting." London: Intellectual Property Research Institute, Queen Mary College, University of London.

———. 2003. "Expanding Intellectual Property's Empire: The Role of FTAs." Geneva: International Centre for Trade and Sustainable Development, November. http://ictsd.org/downloads/2008/08/drahos-fta-2003-en.pdf.

Drahos, Peter, and John Braithwaite. 2002. *Information Feudalism: Who Owns the Knowledge Economy?* London: Earthscan.

———. 2004. "Who Owns the Knowledge Economy? Political Organising behind TRIPS." Lancashire, UK: Corner House.

Dreher, Axel, and Nathan Jensen. 2007. "Independent Actor or Agent? An Empirical Analysis of the Impact of US Interests on IMF Conditions." *Journal of Law and Economics* 50:105–124.

Dreher, Axel, and Jan-Egbert Sturm. 2006. "Do IMF and World Bank Influence Voting in the UN General Assembly?" Zurich: KOF.

Dreher, Axel, Jan-Egbert Sturm, and James Vreeland. 2006. "Does Membership on the UN Security Council Influence IMF Decisions? Evidence from Panel Data." New Haven, CT: Leitner Program in International and Comparative Political Economy.

Dreiling, Michael. 2001. *Solidarity and Contention: The Politics of Security and Sustainability in the NAFTA Conflict*. Edited by S. Bruchey. New York: Garland.

du Gay, Paul, Stuart Hall, Linda Janes, Hugh Mackay, and Keith Negus. 1997. *Doing Cultural Studies: The Story of the Sony Walkman*. Thousand Oaks, CA: Sage.

Dunfield, Kari, and James Germida. 2004. "Impact of Genetically Modified Crops on Soil- and Plant-Associated Microbial Communities." *Journal of Environmental Quality* 33:806–815.

Dymond, Abi. 2007. "Undermining Development? Copper Mining in Zambia." London: Christian Aid.

Earth Rights International. 2011. "The Yadana Pipeline." http://www.earthrights.org/campaigns/yadana-pipeline (accessed May 25, 2011).

Eastham, Katie, and Jeremy Sweet. 2002. "Genetically Modified Organisms (GMOs): The Significance of Gene Flow through Pollen Transfer." Copenhagen: European Environment Agency.

Economist. 2002. "Saddam's Charm Offensive." October 10. http://www.economist.com/node/1378764.

———. 2008. "Explosive Mine: Bougainville." February 7. LexisNexis Academic.

———. 2009. "Cap and Tirade." December 3, 14–15.

Eisenhower Study Group. 2011. "The Costs of War since 2001: Iraq, Afghanistan, and Pakistan." Providence, RI: Watson Institute, Brown University.

Ellerman, A. Denny, and Paul L. Joskow. 2008. "The European Union's Emissions Trading System in Perspective." Arlington, VA: Pew Center on Global Climate Change.

Emmanuel, Arghiri. 1972. Unequal Exchange: A Study of the Imperialism of Trade. New York: Monthly Review Press.

Endres, Alfred, and Volker Radke. 2012. Economics for Environmental Studies: A Strategic Guide to Micro- and Macroeconomics. New York: Springer-Verlag.

Eng, Li Li, and Hean Tat Keh. 2007. "The Effects of Advertising and Brand Value on Future Operating and Market Performance." Journal of Advertising 36:91–100.

EPA (U.S. Environmental Protection Agency). 2004. "Cleaning Up the Nation's Waste Sites: Markets and Technology Trends." Washington, DC: EPA.

———. 2009. "Ways in Which Revisions to the American Clean Energy and Security Act Change the Projected Economic Impacts of the Bill." Washington, DC: EPA.

———. 2010. "2009 Toxics Release Inventory National Analysis Overview." Washington, DC: EPA.

Environment News Service. 2012. "Chevron Faces Midnight Deadline in $19 Billion Ecuador Judgment." August 6. http://ens-newswire.com/2012/08/06/chevron-faces-midnight-deadline-in-19-billion-ecuador-judgment/.

Epstein, Paul, and Jesse Selber. 2002. "Oil: A Life Cycle Analysis of Its Health and Environmental Impacts." Cambridge: Center for Health and the Global Environment, Harvard Medical School.

ETC Group. 2008. "Who Owns Nature? Corporate Power and the Final Frontier in the Commodification of Life." Ottawa: ETC Group.

Eurodad. 2006. "World Bank and IMF Conditionality: A Development Injustice." Brussels: Eurodad.

Evans, Geoff, James Goodman, and Nina Lansbury, eds. 2002. Moving Mountains: Communities Confront Mining and Globalisation. London: Zed Books.

FAO (Food and Agriculture Organization of the United Nations). 2008a. "Current World Fertilizer Trends and Outlook to 2012." Rome: FAO.

———. 2008b. "Enabling Agriculture to Contribute to Climate Change Mitigation." Rome: FAO.

———. 2009. "Feeding the World, Eradicating Hunger: Executive Summary." Rome: FAO.

———. 2010a. "The Role of Agriculture in Climate Change Mitigation." Rome: FAO. Accessed May 8, 2012. http://www.fao.org/es/esa/pesal/AgRole2.html

———. 2010b. "The State of Food Insecurity in the World: Addressing Food Insecurity in Protracted Crises." Rome: FAO.

———. 2011. "Save and Grow: A Policymaker's Guide to the Sustainable Intensification of Smallholder Crop Production." Rome: FAO.

Farago, Robert. 2009. "The Truth about Rare Earths and Hybrids." *International Business Times*, July 23. http://www.ibtimes.com/contents/20090723/editorial-truth-about-rare-earths-and-hybrids.htm.

Farquhar, Iain, and Alistair Smith. 2006. "Collateral Damage: How Price Wars between UK Supermarkets Helped to Destroy Livelihoods in the Banana and Pineapple Supply Chains." Norwich, UK: Banana Link, November.

Farrell, Leanne, Payal Sampat, Radhika Sarin, and Keith Slack. 2004. "Dirty Metals: Mining, Communities and the Environment." Washington, DC: Earthworks and Oxfam America.

Ferguson, Thomas, and Joel Rogers. 1981. "The Reagan Victory: Corporate Coalitions in the 1980 Campaign." In *The Hidden Election: Politics and Economics in the 1980 Presidential Campaign*, edited by Thomas Ferguson and Joel Rogers, 3–64. New York: Pantheon Books.

———. 1986. *Right Turn: The Decline of the Democrats and the Future of American Politics*. New York City: Hill and Wang.

Fields, Scott. 2001. "Tarnishing the Earth: Gold Mining's Dirty Secret." *Environmental Health Perspectives* 109:A474–A481.

Finn, Kathy. 2012. "Two Years after BP Oil Spill Tourists Back in Gulf Coast." Reuters, May 27. http://www.reuters.com/article/2012/05/27/us-usa-bpspill-tourism-idUSBRE84Q09J20120527.

Fiske, Susan, and Shelley Taylor. 1991. *Social Cognition*. New York: McGraw-Hill.

Flanigan, Peter. 1973a. "Agricultural Trade and the Proposed Round of Multilateral Negotiations." Washington, DC: Council on International Economic Policy.

———. 1973b. "International Economic Report of the President Together with the Annual Report of the Council on International Economic Policy." Washington, DC: Council on International Economic Policy.

———. 1974. "International Economic Report of the President Together with the Annual Report of the Council on International Economic Policy." Washington, DC: Council on International Economic Policy.

Fleck, Robert, and Christopher Kilby. 2006. "World Bank Independence: A Model and Statistical Analysis of U.S. Influence." *Review of Development Economics* 10:224–240.

Fligstein, Neil. 2001. *The Architecture of Markets: An Economic Sociology of Twenty-First-Century Capitalist Societies*. Princeton: Princeton University Press.

Foster, John Bellamy. 1994. *The Vulnerable Planet: A Short Economic History of the Environment*. New York: Monthly Review Press.

———. 1999. "Marx's Theory of Metabolic Rift: Classical Foundations for Environmental Sociology." *American Journal of Sociology* 105:366–405.

————. 2008. "Peak Oil and Energy Imperialism." *Monthly Review* 60:12–33.

————. 2012. "The Planetary Rift and the New Human Exemptionalism: A Political-Economic Critique of Ecological Modernization Theory." *Organization & Environment* 25:211–237.

Foster, John Bellamy, Brett Clark, and Richard York. 2010. *The Ecological Rift: Capitalism's War on the Earth*. New York: Monthly Review Press.

Foundation Center. 2001. "The Foundation Grants Index on CD-ROM." New York: Foundation Center.

————. 2002. "The Foundation Grants Index on CD-ROM." New York: Foundation Center.

Foust, Joshua. 2112. "Counting Afghanistan's Dead." *Atlantic*, July 20. http://www.theatlantic.com/international/archive/2012/07/counting-afghanistans-dead/260053/.

Frank, Thomas. 1997. *Conquest of Cool: Business Culture, Counterculture, and the Rise of Hip Consumerism*. Chicago: University of Chicago Press.

Freeport McMoRan Copper & Gold, Inc. 2007. "2007 Annual Report: A World of Assets A World of Opportunities." Phoenix, AZ: Freeport-McMoRan Copper & Gold, Inc.

Freudenburg, William, and Margarita Alario. 2007. "Weapons of Mass Distraction: Magicianship, Misdirection, and the Dark Side of Legitimation." *Sociological Forum* 22:146–173.

Gaventa, John. 1980. *Power and Powerlessness: Quiescence and Rebellion in an Appalachian Valley*. Urbana: University of Illinois Press.

Gedicks, Al. 2001. *Resource Rebels: Naive Challenges to Mining and Oil Corporations*. Cambridge, MA: South End.

Gereffi, Gary, John Humphrey, and Timothy Sturgeon. 2005. "The Governance of Global Value Chains." *Review of International Political Economy* 12:78–104.

Gibney, Mark, Linda Cornet, and Reed Wood. 2009. "About the Political Terror Scale." Political Terror Scale. http://www.politicalterrorscale.org/about.php (accessed March 1, 2009).

Gigerenzer, Gird, and Reinhard Selten. 2001. *Bounded Rationality: The Adaptive Toolbox*. Cambridge: MIT Press.

Gillibrand, Kirsten. 2009. "Cap and Trade Could Be a Boon to New York: The City Is Uniquely Positioned to Benefit from a Global Market in Carbon Emissions." *Wall Street Journal*, October 21. http://online.wsj.com/article/SB10001424052748704500604574481812686144826.html#articleTabs%3Darticle.

Gold, Joseph. 1988. "Mexico and the Development of the Practice of the International Monetary Fund." *World Development* 16:1127–1142.

Goldman, Michael. 2005. *Imperial Nature: The World Bank and Struggles for Social Justice in the Age of Globalization*. Edited by J. C. Scott. New Haven: Yale University Press.

Goodland, Robert, and Jeff Anhang. 2009. "Livestock and Climate Change: What If the Key Actors in Climate Change Are . . . Cows, Pigs, and Chickens?" Washington, DC: Worldwatch Institute.

Goodstein, Eban. 2002. *Economics and the Environment*. New York: Wiley.

Gordon, David M., Richard Edwards, and Michael Reich. 1982. *Segmented Work, Divided Workers*. New York: Cambridge University Press.

Gore, Al. 2009. *Our Choice: A Plan to Solve the Climate Crisis*. Emmaus, PA: Rodale.

Gorelick, Steven. 2000. "The Farm Crisis: How We Are Killing the Small Farmer." Berkeley, CA: International Society for Ecology and Culture.

Gould, Kenneth. 1994. "Legitimacy and Growth in the Balance: The Role of the State in Environmental Remediation." *Industrial & Environmental Crisis Quarterly* 8:237–256.

Gould, Kenneth A., David N. Pellow, and Allan Schnaiberg. 2004. "Interrogating the Treadmill of Production." *Organization & Environment* 17:296–316.

———. 2008. *The Treadmill of Production: Injustice and Unsustainability in the Global Economy*. Boulder, CO: Paradigm.

Gould, Kenneth, Adam Weinberg, and Allan Schnaiberg. 1995. "Natural Resource Use in a Transnational Treadmill: International Agreements, National Citizenship Practices, and Sustainable Development." *Humboldt Journal of Social Relations* 21:60–93.

GRAIN. 2008. "Seized! The 2008 Land Grab for Food and Financial Security." Barcelona: GRAIN, October 24. http://www.grain.org/article/entries/93-seized-the-2008-landgrab-for-food-and-financial-security.

Green, Jonathan. 2007. "The Biofuel Time Bomb." *Daily Mail*, April 28. LexisNexis Academic.

Green Guide magazine editors. 2008. *Green Guide: The Complete Reference for Consuming Wisely*. Washington, DC: National Geographic Ventures.

Grube, Arthur, David Donaldson, Timothy Kiely, and La Wu. 2011. "Pesticides Industry Sales and Usage: 2006 and 2007 Market Estimates." Washington, DC: U.S. Environmental Protection Agency.

Gupta, Akhil. 2012. *Red Tape: Bureaucracy, Structural Violence, and Poverty in India*. Durham: Duke University Press.

Gurian-Sherman, Doug. 2008. "CAFOs Uncovered: The Untold Costs of Confined Animal Feeding Operations." Cambridge, MA: Union of Concerned Scientists.

———. 2009. "Failure to Yield: Evaluating the Performance of Genetically Engineered Crops." Cambridge, MA: Union of Concerned Scientists.

Gurr, Geoff, Steve Wratten, and Miguel Altieri. 2004. *Ecological Engineering for Pest Management: Advances in Habitat Manipulation for Anthropods*. Ithaca: Cornell University Press.

Gwin, Catherine. 1997. "U.S. Relations with the World Bank, 1945–1992." In *The World Bank: Its First Half Century*, vol. 2, edited by Devesh Kapur, John P. Lewis, and Richard Webb, 195–274. Washington, DC: Brookings Institution Press.

Halford, Jason, Emma Boyland, Georgina Hughes, Lorraine Oliveira, and Terence Dovey. 2007. "Beyond-Brand Effect of Television (TV) Food Advertisements/Commercials on Caloric Intake and Food Choice of 5–7-Year-Old Children." *Appetite* 49:263–267.

Hall, Anthony. 1989. *Developing Amazonia: Deforestation and Social Conflict in Brazil's Carajas Programme*. Manchester: Manchester University Press.

Hanjra, Munir A., and M. Ejaz Qureshi. 2010. "Global Water Crisis and Future Food Security in an Era of Climate Change." *Food Policy* 35:365–377.

Hanley, Nick, Jason Shogren, and Ben White. 2013. *Introduction to Environmental Economics*. New York: Oxford University Press.

Harden, Blaine. 2001. "The Dirt in the New Machine." *New York Times Magazine*, August 12, 35.

Harrigan, Jane, Chengang Wang, and Hamed El-Said. 2006. "The Politics of IMF and World Bank Lending: Will It Backfire in the Middle East and North Africa?" In *The IMF, World Bank, and Policy Reform*, edited by Alberto Paloni and Maurizio Zanardi, 64–99. New York: Routledge.

Harris, Jennifer, John Bargh, and Kelly Brownell. 2009. "Priming Effects of Television Food Advertising on Eating Behavior." *Health Psychology* 28:404–413.

Harrison, Graham. 2004. *The World Bank and Africa: The Construction of Governance States*. New York: Routledge.

Harrison, Jill. 2011. *Pesticide Drift and the Pursuit of Environmental Justice*. Cambridge: MIT Press.

Hartung, William, and Michelle Ciarrocca. 2003. "The Military-Industrial-Think Tank Complex." *Multinational Monitor* 24 (1–2). http://multinationalmonitor.org/mm2003/03jan-feb/jan-feb03corp2.html.

Haslett, Malcolm. 2001. "Afghanistan: The Pipeline War?" BBC News, October 29. http://news.bbc.co.uk/2/hi/south_asia/1626889.stm.

Hassan, Ghali. 2005. "Living Conditions in Iraq: A Criminal Tragedy." Centre for Research on Globalisation, June 3. http://www.globalresearch.ca/articles/HAS506A.html.

Hauter, Wenonah. 2009. "Agriculture and Antitrust Enforcement Issues in Our 21st Century Economy." Letter submitted to the U.S. Department of Justice and U.S. Department of Agriculture on Agriculture and Antitrust Enforcement Issues in Our 21st Century Economy (74 Fed. Reg. 165, 43725-43726), December 31. Washington, DC: Water Watch Institute.

———. 2012. *Foodopoly: The Battle over the Future of Food and Farming in America*. New York: New Press.

Hawken, Paul, Amory Lovins, and L. Hunter Lovins. 1999. *Natural Capitalism: Creating the Next Industrial Revolution*. New York: Little, Brown.

Hayes, Lynn. 2009. "Farmers' Guide to GMOs." St. Paul, MN: Farmers' Legal Action Group.

Hearn, Kelly 2006. "Exclusive: Selling the Amazon for a Handful of Beads." *AlterNet*, January 16. http://www.alternet.org/story/30657.

Heffernan, William. 2000. "Concentration of Ownership and Control in Agriculture." In *Hungry for Profit: The Agribusiness Threat to Farmers, Food, and the Environment*, edited by Fred Magdoff, John Bellamy Foster, and Frederick H. Buttel, 61–76. New York: Monthly Review Press.

Heffernan, William, Mary Hendrickson, and Robert Gronski. 1999. "Consolidation in the Food and Agriculture System." Columbia: Department of Rural Sociology, University of Missouri.

Hendrickson, Mary, and William Heffernan. 2007. "Concentration of Agricultural Markets: April 2007." Columbia: Department of Rural Sociology, University of Missouri.

Hendrickson, Mary, John Wilkinson, William Heffernan, and Robert Gronski. 2008. "The Global Food System and Nodes of Power." Boston: Oxfam America.

Herald Sun. 2005. "BHP Mine Stays Closed." May 27. LexisNexis Academic.

Herr, Norman. 2007. "Television & Health." Norman Herr's California State University, Northridge, website. http://www.csun.edu/science/health/docs/tv&health. html#tv_stats.

Hershey, Robert. 1986. "Shifts at Enterprise Institute." *New York Times,* June 27. http:// www.nytimes.com/1986/06/27/business/shifts-at-enterprise-institute.html.

Hilson, Gavin. 2004. "Structural Adjustment in Ghana: Assessing the Impacts of Mining-Sector Reform." *Africa Today* 51:53–77.

Himmelstein, Jerome. 1990. *To the Right: The Transformation of American Conservatism.* Berkeley: University of California Press.

Hoffman, Andy. 2007. "Out of Myanmar, but the Cheques Keep Coming." *Globe and Mail,* October 3, B1.

Holguin, Jaime. 2009. "'Shock and Awe' Throttles Iraq." CBS News, March 22. http:// www.cbsnews.com/stories/2003/03/22/iraq/main545267.shtml.

Hooks, Gregory, and Chad L. Smith. 2004. "The Treadmill of Destruction: National Sacrifice Areas and Native Americans." *American Sociological Review* 69:558–575.

———. 2005. "Treadmills of Production and Destruction." *Organization & Environment* 18:19–37.

Horton, Amy. 2010. "World Bank Voting Reforms: Governance Remains Illegitimate." London: Bretton Woods Project.

Hossein-zadeh, Ismael 2006. "Behind the Plan to Bomb Iran." *Payvand Iran News,* August 30. http://www.payvand.com/news/06/aug/1332.html.

House Committee on Science and Technology, Subcommittee on Natural Resources, Agriculture Research, and Environment. 1984. *Tropical Forest Development Projects: Status of Environmental and Agricultural Research.* Washington, DC: U.S. Government Printing Office.

Howard, Philip H. 2009. "Visualizing Consolidation in the Global Seed Industry: 1996–2008." *Sustainability* 1:1266–1287.

Hubbard, Kristina. 2009. "Out of Hand: Farmers Face the Consequences of a Consolidated Seed Industry." Washington, DC: National Family Farm Coalition.

Human Rights Watch. 1999. "The Price of Oil: Corporate Responsibility and Human Rights Violations in Nigeria's Oil Producing Communities." New York: Human Rights Watch.

Humphreys, David. 2006. *Logjam: Deforestation and the Crisis of Global Governance.* London: Earthscan.

Iallonard, Tony. 2010. "BP Oil Spill Draws Thousands of Concerned Sportsmen to Virtual Town Hall." National Wildlife Federation, May 6. http://www.nwf.org/ News-and-Magazines/Media-Center/News-by-Topic/Wildlife/2010/05-06-10-BP-Oil-Spill-Draws-Thousands-of-Concerned-Sportsmen-to-Virtual-Town-Hall.aspx.

iCasualties. 2012. "Operation Enduring Freedom: Coalition Military Fatalities by Year"; "Operation Iraqi Freedom: Coalition Military Fatalities by Year." http://icasualties. org/oef/; http://icasualties.org/Iraq/index.aspx (accessed October 24, 2012).

ICTSD (International Centre for Trade and Sustainable Development). 2009. "Agricultural Subsidies in the WTO Green Box: Ensuring Coherence with Sustainable Development Goals." Geneva: ICTSD, September. http://ictsd.org/downloads/2012/02/agricultural-subsidies-in-the-wto-green-box-ensuring-coherence-with-sustainable-development-goals.pdf.

Igbikiowubo, Hector. 2004. "US Offers Nigeria Military Aid to Protect Offshore Oil." All Africa, July 25. http://www.energybulletin.net/node/1211.

IMF (International Monetary Fund). 2008. "How the IMF Helps to Resolve Balance of Payments Difficulties." 2011. Washington, DC: IMF.

———. 2011. "Quota and Voting Shares before and after Implementation of Reforms Agreed in 2008 and 2010." Washington, DC: IMF. http://www.imf.org/external/np/sec/pr/2011/pdfs/quota_tbl.pdf.

Independent Evaluation Office of the IMF. 2007. "Aspects of IMF Corporate Governance—Including the Role of the Board." Washington, DC: IMF.

Institute of Medicine. 2011. "Long-Term Health Consequences of Exposure to Burn Pits in Iraq and Afghanistan." Washington, DC: National Academy of Sciences.

International Energy Agency. 2012. World Energy Outlook 2012: Executive Summary. Paris: Organization for Economic Co-operation and Development.

International Rivers. 2008. "Nam Theun 2 Hydropower Project: Risky Business for Laos." Berkeley, CA: International Rivers.

———. 2009. "Mekong Mainstream Dams: Threatening Southeast Asia's Food Security." Berkeley, CA: International Rivers.

———. 2010. "Nam Theun 2 Hydropower Project: The Real Cost of a Controversial Dam." Berkeley, CA: International Rivers.

International Trade Centre. 2011. "The Impacts of Private Standards on Global Value Chains: Literature Review Series on the Impacts of Private Standards—Part I." Geneva: International Trade Centre.

IPCC (Intergovernmental Panel on Climate Change). 2007. "Climate Change 2007: Synthesis Report." Geneva: IPCC.

Iraq Body Count. 2012. "Iraq Body Count." http://www.iraqbodycount.org/ (accessed October 26, 2012).

IRIN. 2006. "Iraq: Unemployment and Violence Increase Poverty." October 17. http://www.irinnews.org/report/61892/iraq-unemployment-and-violence-increase-poverty.

Island Press. 1999. "The Environmental Impacts of War." Eco-Compass. http://www.islandpress.org/ecocompass/war/war.html.

Jaffee, Daniel. 2007. Brewing Justice: Fair Trade Coffee, Sustainability, and Survival. Berkeley: University of California Press.

Jamail, Dahr. 2007. "Iraq's Forgotten Refugees." Alternet, April 23. http://www.alternet.org/world/50946/?page=entire.

———. 2011. "Stress and Anger over BP Oil Disaster Could Linger for Decades." Inter Press Service News Agency, April 15. http://www.ipsnews.net/2011/04/stress-and-anger-over-bp-oil-disaster-could-linger-for-decades/.

———. 2012. "Gulf Fisheries in Decline after Oil Disaster." Al Jazeera, April 19. http://www.aljazeera.com/indepth/features/2012/03/20123571723894800.html.

James, Clive. 2011. "Global Status of Commercialized Biotech/GM Crops: 2011; ISAAA Brief No. 43." Ithaca, NY: International Service for the Acquisition of Agri-Biotech Applications.

James, Harvey, Mary Hendrickson, and Philip Howard. 2012. "Networks, Power and Dependency in the Agrifood Industry." Columbia: College of Agriculture, Food & Natural Resources, University of Missouri.

James, Steven. 2006. "Islanders Win Appeal in Claim against Rio Tinto." Reuters, August 8. http://www.planetark.com/dailynewsstory.cfm/newsid/37553/story.htm.

Jawara, Fatoumata, and Aileen Kwa. 2003. *Behind the Scenes at the WTO: The Real World of International Trade Negotiations.* New York: Zed Books.

Jigang, Zhou, and Zhu Chuhua. 2008. "In China's Mining Region, Villagers Stand Up to Pollution." Yale Environment 360, December 4. http://e360.yale.edu/content/feature.msp?id=2095.

Johnson, Allen, Jr., Laurel Calkins, and Margaret Cronin Fisk. 2012. "BP Spill Victims Face Economic Fallout Two Years Later." Bloomberg, February 23. http://www.bloomberg.com/news/2012-02-23/bp-oil-spill-haunts-gulf-business-owners-almost-two-years-after-disaster.html.

Johnson, Bradley. 2010. "Top 100 Outlays Plunge 10% but Defying Spend Trend Can Pay Off." *Advertising Age* 81 (25): 1, 10–11.

Johnson, Chalmers. 2004. *The Sorrows of Empire.* New York: Metropolitan Books.

Jordan, Jillian. 2012. "America's Military Renewables Plan Fast-Tracked and Mission Critical." *Energy Exchange* (blog), Environmental Defense Fund, August 23. http://blogs.edf.org/energyexchange/2012/08/23/americas-military-renewables-plan-fast-tracked-and-mission-critical/.

Jorgenson, Andrew. 2007. "Does Foreign Investment Harm the Air We Breathe and the Water We Drink? A Cross-National Study of Carbon Dioxide Emissions and Organic Water Pollution in Less-Developed Countries, 1975–2000." *Organization & Environment* 20:137–156.

———. 2009. "The Sociology of Unequal Exchange in Ecological Context: A Panel Study of Lower-Income Countries, 1975–2000." *Sociological Forum* 24:22–46.

Jorgenson, Andrew, and Thomas. Burns. 2007. "The Political-Economic Causes of Change in the Ecological Footprints of Nations, 1991–2001: A Quantitative Investigation." *Social Science Research* 36:834–853.

Jorgenson, Andrew, and Brett Clark. 2009. "The Economy, Military, and Ecologically Unequal Relationships in Comparative Perspective: A Panel Study of the Ecological Footprints of Nations, 1975–2000." *Social Problems* 56:621–646.

Jorgenson, Andrew, Brett Clark, and Jeffrey Kentor. 2010. "Militarization and the Environment: A Panel Study of Carbon Dioxide Emissions and the Ecological Footprints of Nations, 1970–2000." *Global Environmental Politics* 10:7–29.

Jorgenson, Andrew K., and Kennon A. Kuykendall. 2008. "Globalization, Foreign Investment Dependence, and Agriculture Production: A Cross-National Study of

Pesticide and Fertilizer Use Intensity in Less-Developed Countries, 1990–2000." *Social Forces* 87:529–560.

Joseph, A., S. Abraham, J. P. Muliyil, K. George, J. Prasad, S. Minz, V. J. Abraham, and K. S. Jacob. 2003. "Evaluation of Suicide Rates in Rural India Using Verbal Autopsies, 1994–9." *British Medical Journal* 326:1121–1122.

Jung, Chulho, and Barry J. Seldon. 1995. "The Macroeconomic Relationship between Advertising and Consumption." *Southern Economic Journal* 61:577–587.

Kahneman, Daniel. 2003. "Maps of Bounded Rationality: Psychology for Behavioral Economics." *American Economic Review* 93:1449–1475.

Kapur, Devesh, John Lewis, and Richard Webb. 1997. *The World Bank: Its First Half Century*. Vol. 1. Washington, DC: Brookings Institution Press.

Kay, Katty. 2001. "Analysis: Oil and the Bush Cabinet." BBC News, January 29. http://news.bbc.co.uk/2/hi/americas/1138009.stm.

Keane, Julian. 2012. "Mozambique Town Transformed by Coal Rush." BBC News, October 21. http://www.bbc.co.uk/news/world-africa-20024326.

Kearney, Marianne. 2006. "Papua Rage Flares—Protests over Mine Turn Deadly." *Courier Mail*, March 18. LexisNexis Academic.

Keck, Margaret E. 1998. "Planafloro in Rondonia: The Limits of Leverage." In *The Struggle for Accountability: The World Bank, NGOs, and Grassroots Movements*, edited by Jonathan A. Fox and L. David Brown, 181–218. Cambridge: MIT Press.

Kennedy, Gina, Guy Nantel, and Prakash Shetty. 2004. "Globalization of Food Systems in Developing Countries: A Synthesis of Country Case Studies." In *Globalization of Food Systems in Developing Countries: Impact on Food Security and Nutrition*, edited by Food and Agriculture Organization, 1–25. Rome: Food and Agriculture Organization.

Kennedy, Kelly. 2008a. "Burn Pit at Balad Raises Health Concerns." *Army Times*, October 27. http://www.armytimes.com/news/2008/10/military_burnpit_102708w/.

———. 2008b. "Report: Army Making Toxic Mess in War Zones." *Army Times*, October 2. http://www.militarytimes.com/news/2008/10/military_toxiciraq_100208w/.

Kentor, Jeffrey, and Peter Grimes. 2006. "Foreign Investment Dependence and the Environment: A Global Perspective." In *Globalization and the Environment*, edited by Andrew Jorgenson and Edward Kick, 68–87. Boston: Brill Academic.

Keteyian, Armen. 2011. "Oil & Gas Industry Spills Happen 'All the Time.'" CBS News, April 12. http://www.cbsnews.com/stories/2011/04/12/eveningnews/main20053283.shtml?tag=contentBody;featuredPost-PE.

Khan, Shahbaz, and Munir A. Hanjra. 2009. "Footprints of Water and Energy Inputs in Food Production—Global Perspectives." *Food Policy* 34:130–140.

Kharaka, Yousif K., and James K. Otton. 2003. "Environmental Impacts of Petroleum Production: Initial Results from the Osage-Skiatook Petroleum Environmental Research Sites, Osage County, Oklahoma." Menlo Park, CA: U.S. Geological Survey.

Kilby, Christopher. 2009. "The Political Economy of Conditionality: An Empirical Analysis of World Bank Loan Disbursements." *Journal of Development Economics* 89:51–61.

Killick, Tony. 1995. *IMF Programs in Developing Countries: Design and Impact*. London: Routledge.

Kirk, Ronald. 2011. "2011 Trade Policy Agenda and 2010 Annual Report of the President of the United States on the Trade Agreements Program." Washington, DC: Office of the United States Trade Representative.

Kirzner, Israel. 1999. "Mises and His Understanding of the Capitalist System." *Cato Journal* 19:215–232.

Klare, Michael. 2001. *Resource Wars: The New Landscape of Global Conflict*. New York: Holt.

———. 2004. *Blood and Oil: The Dangers and Consequences of America's Growing Dependency on Imported Petroleum*. New York: Holt.

Knudson, Tom. 2009. "The Cost of Biofuel: Destroying Indonesia's Forests." Yale Environment 360, January 19. http://e360.yale.edu/content/feature.msp?id=2112.

Kobb, Kurt. 2013. "Will the International Energy Agency's Oil Forecast Be Wrong Again." *Resource Insights* (blog), May 19. http://www.resilience.org/stories/2013-05-19/will-the-international-energy-agency-s-oil-forecast-be-wrong-again.

Kolko, Joyce. 1988. *Restructuring the World Economy*. New York: Pantheon Books.

Korten, David. 2001. *When Corporations Ruled the World*. Bloomfield, CT: Kumarian.

Kovalik, Daniel. 2010. "Obama's War for Oil in Colombia." *Counterpunch*, January 27. http://www.counterpunch.org/2010/01/27/obama-s-war-for-oil-in-colombia/.

Kramer, Andrew. 2011. "In Rebuilding Iraq's Oil Industry, U.S. Subcontractors Hold Sway." *New York Times*, June 17, B1, B8.

Krebs, A. V. 1999. "Cargill & Co.'s Comparative Advantage in Free Trade." *Agribusiness Examiner*, April 26. http://www.electricarrow.com/carp/agbiz/agex-31.html.

Krehely, Jeff, Meaghan House, and Emily Kernan. 2004. "Axis of Ideology: Conservative Foundations and Public Policy." Washington, DC: National Committee for Responsive Philanthropy.

Kumar, Claire. 2009. "Undermining the Poor: Mineral Taxation Reforms in Latin America." London: Christian Aid.

Kupfer, David. 1996. "Worldwide Shell Boycott—Shell Oil; Nigeria Executes Niger Delta Environmental Protester/Writer Ken Saro-Wiwa and 8 Others." *Progressive*, January. http://findarticles.com/p/articles/mi_m1295/is_n1_v60/ai_17963624/.

Kwa, Aileen. 2002. "Power Politics in the WTO: Developing Countries' Perspectives on Decision-Making Processes in Trade Negotiations." Bangkok: Focus on the Global South.

Lal, Deepak. 2005. *In Defense of Empires*. Washington, DC: American Enterprise Institute.

Lambrechts, Kato. 2009. "Breaking the Curse: How Transparent Taxation and Fair Taxes Can Turn Africa's Mineral Wealth into Development." Johannesburg, South Africa: Open Society Institute of Southern Africa, Third World Network, Tax Justice Africa, Action Aid International, and Christian Aid.

Langston, Jennifer. 2004. "Quest for Justice on Bougainville Continues." *Seattle Post-Intelligencer*, July 19. http://www.minesandcommunities.org/article.php?a=680.

Lappe, Anna. 2010. *Diet for a Hot Planet: The Climate Crisis at the End of Your Fork and What You Can Do about It.* New York: Bloomsbury.

Larmer, Brook. 2009. "The Real Price of Gold." *National Geographic*, January. http:// ngm.nationalgeographic.com/print/2009/01/gold/larmer-text.

La Rovere, Emilio L., and Francisco E. Mendes. 2000. *Tucurui Hydropower Complex Brazil.* Cape Town, South Africa: World Commission on Dams.

LaSalle, Tim, and Paul Hepperly. 2008. "Regenerative Organic Farming: A Solution to Global Warming." Kutztown, PA: Rodale Institute.

Lawrence, Shannon. 2008a. "Nam Theun 2 Dam: Rising Water, Falling Expectations." Berkeley, CA: International Rivers.

———. 2008b. "Nam Theun 2: Trip Report and Project Update." Berkeley, CA: International Rivers.

Lehner, Peter. 2011. "Military Top Brass Support Clean Energy Development: It's a Matter of National Security." *Switchboard: Natural Resources Defense Council Staff Blog*, October 11. http://switchboard.nrdc.org/blogs/plehner/military_top_brass_support_cle.html.

Leith, Denise. 2003. *The Politics of Power: Freeport in Suharto's Indonesia.* Honolulu: University of Hawai'i Press.

Lele, Uma, Nalini Kumar, Syed Husain, Aaron Azaueta, and Lauren Kelly. 2000. "The World Bank Forest Strategy: Striking the Right Balance." Washington, DC: World Bank Operations Evaluation Department.

Lengyel, Gregory J. 2007. "Department of Defense Energy Strategy: Teaching an Old Dog New Tricks." Washington, DC: Brookings Institution.

Lewis, Randall, and David Reiley. 2009. "Retail Advertising Works! Measuring the Effects of Advertising on Sales via a Controlled Experiment on Yahoo!" Sunnyvale, CA: Yahoo Research Technical Report.

Lohmann, Larry. 2001. "Democracy or Carbocracy? Intellectual Corruption and the Future of the Climate Debate." Dorset, UK: Corner House.

Lokupitya, Erandathie, and Keith Paustian. 2006. "Agricultural Soil Greenhouse Gas Emissions: A Review of National Inventory Methods." *Journal of Environmental Quality* 35:1413–1427.

Lutz, Catherine. 2013. "US and Coalition Casualties in Iraq and Afghanistan." Providence, RI: Watson Institute for International Studies, Brown University.

MacDonald, David. 2009. *Thinking History, Fighting Evil: Neoconservatives and the Perils of Analogy in American Politics.* Boulder, CO: Lexington Books.

MacDonald, James, and Penni Korb. 2008. "Agricultural Contracting Update, 2005." Washington, DC: U.S. Department of Agriculture.

———. 2011. "Agricultural Contracting Update: Contracts in 2008." Washington, DC: U.S. Department of Agriculture.

MacDonald, James, Janet Perry, Mary Ahearn, David Banker, William Chambers, Carolyn Dimitri, Nigel Key, Kenneth Nelson, and Leland Southard. 2004. "Contracts, Markets, and Prices: Organizing the Production and Use of Agricultural Commodities." Washington, DC: U.S. Department of Agriculture.

Machlis, Gary, and Thor Hanson. 2008. "Warfare Ecology." *BioScience* 58:729–736.

MacKinnon, Ian. 2007. "Civil War: Burma's Bloody Battle for Power: Villagers Killed, Homes Destroyed as Junta Seeks to Control Natural Resources." *Guardian*, May 15. LexisNexis Academic.

Mann, James. 2004. *The Rise of the Vulcans: The History of Bush's War Cabinet*. New York: Penguin.

Mann, Michael. 1986. *The Sources of Social Power: A History of Power from the Beginning to A.D. 1760*. Cambridge: Cambridge University Press.

Markham, Victoria, and Nadia Steinzor. 2006. "U.S. National Report on Population and the Environment." New Canaan, CT: Center for Environment and Population.

May, Christopher. 2007. *The World Intellectual Property Organization: Resurgence and the Development Agenda*. New York: Routledge.

Mayer, Jane. 2004. "Contract Sport: What Did the Vice-President Do for Halliburton?" *New Yorker*, February 16. http://www.newyorker.com/archive/2004/02/16/040216fa_fact.

McCright, Aaron, and Riley Dunlap. 2003. "Defeating Kyoto: The Conservative Movement's Impact on U.S. Climate Change Policy." *Social Problems* 50:348–373.

McDowell, David, Thayer Scudder, and Lee M. Talbot. 2008. "Environmental and Social Monitoring Report: LAO: Greater Mekong Subregion: Nam Theun 2 Hydroelectric." Mandaluyong City, Philippines: Asian Development Bank.

———. 2009. "Fifteenth Report of the International Environmental and Social Panel of Experts for the Nam Theun 2 Multipurpose Project Lao People's Democratic Republic." Washington DC: World Bank.

McMahon, Gary. 2010. "The World Bank's Evolutionary Approach to Mining Sector Reform." Washington, DC: World Bank.

McMichael, Philip. 2008. *Development and Social Change: A Global Perspective*. Los Angeles: Pine Forge.

McNeill, J. R., and David S. Painter. 2009. "The Global Environmental Footprint of the U.S. Military 1789–2003." In *War and the Environment: Military Destruction in the Modern Age*, edited by Charles E. Closmann, 10–31. College Station: Texas A&M University Press.

McQuarrie, Edward, and Barbara Phillips. 2005. "Indirect Persuasion in Advertising: How Consumers Process Metaphors Presented in Picture and Words." *Journal of Advertising* 34:7–20.

Media Transparency. 2007. "Media Transparency: The Money behind the Conservative Media." http://www.mediatransparency.org.

Mekong River Commission. 2011a. "About the MRC." http://www.mrcmekong.org/about-mrc.htm.

———. 2011b. "About the MRC: The Land & Its Resources." http://www.mrcmekong.org/about_mekong/about_mekong.htm.

Melzter, Allan. 2000. "International Financial Institution Advisory Commission Final Report." Washington, DC: International Financial Institution Advisory Commission.

Meo, Nick. 2007. "Rumble in the Jungle." *South China Morning Post*, October 29. Lexis-Nexis Academic.

Mermin, Jonathon. 1999. *Debating War and Peace: Media Coverage of U.S. Intervention in the Post-Vietnam Era.* Princeton: Princeton University Press.

Mikkelsen, Caecilie. 2013. *The Indigenous World 2013.* Edison, NJ: Transaction.

Mines and Communities. 2007. "Rio Tinto Update." August 22. http://www.minesand-communities.org/article.php?a=684.

Mining Journal. 2000. "Mining Annual Review 2000." http://www.fdi.net/documents/WorldBank/conferences/mining2000/Africadata/contents.pdf.

Mining Magazine. 1993. "Manganese Mining Project Update." May, 323.

Mining Watch. 2005. "Sulawesi Communities Reject Inco, Call for Renegotia-tion of PT Inco's Contract of Work." August 12. http://www.miningwatch.ca/sulawesi-communities-reject-inco-call-renegotiation-pt-incos-contract-work.

———. 2006. "Inco's Goro Site Blockaded Again by Indigenous Kanaks: French Troops Respond with Violent Action." April 3. http://www.miningwatch.ca/inco-s-goro-site-blockaded-again-indigenous-kanaks-french-troops-respond-with-violent-action.

Mizruchi, Mark. 1992. *The Structure of Corporate Political Action: Interfirm Relations and Their Consequences.* Cambridge: Harvard University Press.

———. 1996. "What Do Interlocks Do? An Analysis, Critique, and Assessment of Research on Interlocking Directorates." *Annual Review of Sociology* 22:271–298.

Mol, Arthur P. J. 1995. *The Refinement of Production: Ecological Modernization Theory and the Chemical Industry.* Utrecht, Netherlands: International Books.

———. 1997. "Ecological Modernization: Industrial Transformations and Environmental Reform." In *The International Handbook of Environmental Sociology*, edited by Michael R. Redclift and Graham Woodgate, 138–149. Northampton, MA: Edward Elgar.

Monast, Jonas, Jon Anda, and Tim Profeta. 2009. "U.S. Carbon Market Design: Regulating Emission Allowances as Financial Instruments." Durham: Nicholas Institute for Environmental Policy Solutions, Duke University.

Monsanto. 1998. "Cargill and Monsanto Announce Global Feed And Processing Biotechnology Joint Venture." Press release, May 14. http://www.prnewswire.com/news-releases/cargill-and-monsanto-announce-global-feed-and-processing-biotechnology-joint-venture-77897552.html.

Moody, Roger. 2005. *The Risks We Run: Mining, Communities and Political Risk Insurance.* Utrecht, Netherlands: International Books.

———. 2007. *Rocks and Hard Places: The Globalization of Mining.* New York: Zed Books.

Moore, Jason W. 2000. "Environmental Crises and the Metabolic Rift in World-Historical Perspective." *Organization & Environment* 13:123–157.

Moss, Diana. 2011. "Competition and Transgenic Seed Systems." *Antitrust Bulletin* 56:81–102.

Multinational Monitor. 2001. "Bush's Corporate Cabinet." May 1. http://multinational-monitor.org/mm2001/01may/may01bushcc.html#ann.

Muravchik, Joshua. 1991. "At Last, Pax Americana." *New York Times*, January 24. http://www.nytimes.com/1991/01/24/opinion/at-last-pax-americana.html.

Murgai, R., M. Ali, and D. Byerlee. 2001. "Productivity Growth and Sustainability in Post Green-Revolution Agriculture: The Case of Indian and Pakistani Punjabs." *World Bank Research Observer* 16:199–218.

Murphy, Sophia. 2002. "Managing the Invisible Hand: Markets, Farmers and International Trade." Minneapolis, MN: Institute for Agriculture and Trade Policy.

———. 2009. "Free Trade in Agriculture: A Bad Idea Whose Time Is Done." *Monthly Review* 61 (3). http://monthlyreview.org/2009/07/01/free-trade-in-agriculture-a-bad-idea-whose-time-is-done.

Murphy, Sophia, Ben Lilliston, and Mary Beth Lake. 2005. "WTO Agreement on Agriculture: A Decade of Dumping." Minneapolis, MN: Institute for Agriculture and Trade Policy.

Nadal, Alejandro, and Timothy Wise. 2004. "The Environmental Costs of Agricultural Trade Liberalization: U.S.-Mexico Maize Trade under NAFTA." In *Globalization and the Environment: Lessons from the Americas*, edited by Liane Schalatek, 29–32. Washington, DC: Heinrich Böll Foundation North America.

Nader, Ralph, and Toby Heaps. 2008. "We Need a Global Carbon Tax: The Cap-and-Trade Approach Won't Stop Global Warming." *Wall Street Journal*, December 3. http://online.wsj.com/article/SB122826696217574539.html.

Naiman, Robert, and Neil Watkins. 1999. "A Survey of the Impacts of IMF Structural Adjustment in Africa: Growth, Social Spending, and Debt Relief." Washington DC: Center for Economic and Policy Research.

Naito, Koh, Felix Remy, and John P. Williams. 2001. *Review of Legal and Fiscal Frameworks for Exploration and Mining*. London: Mining Journal Books.

Narlikar, Amrita. 2001. "WTO Decision-Making and Developing Countries." Geneva: South Centre.

National Academy of Agricultural Sciences. 2006. "WTO and Indian Agriculture: Implications for Policy and R&D, Policy Paper 38." New Delhi: National Academy of Agricultural Sciences.

National Wildlife Federation. 2015. "How Does the BP Oil Spill Impact Wildlife and Habitat." http://www.nwf.org/What-We-Do/Protect-Habitat/Gulf-Restoration/Oil-Spill/Effects-on-Wildlife.aspx (accessed March 23, 2015).

Nelson, Gerald C., Mark W. Rosegrant, Jawoo Koo, Richard Robertson, Timothy Sulser, Tingju Zhu, Claudia Ringler, Siwa Msangi, Amanda Palazzo, Miroslav Batka, Marilia Magalhaes, Rowena Valmonte-Santos, Mandy Ewing, and David Lee. 2009. "Climate Change: Impact on Agriculture and Costs of Adaptation." Washington, DC: International Food Policy Research Institute.

NEPDG (White House National Energy Policy Development Group). 2001. "National Energy Policy: Reliable, Affordable, and Environmentally Sound Energy for America's Future." Washington, DC: NEPDG.

New York Times. 1992. "Excerpts from Pentagon's Plan: 'Prevent the Re-Emergence of a New Rival.'" March 8, A14.

———. 2006. "After Clashes, Indonesian Troops Guard Gold Mine." February 25. Lexis-Nexis Academic.

———. 2007. "More Isolated Indians Survive in Amazon Rain Forest, but Face Peril." January 18. LexisNexis Academic.

Noam, Eli M. 2009. *Media Ownership and Concentration in America*. New York City: Oxford University Press.

Noble-Wilford, John. 2008. "Twilight for the Forest People." *New York Times*, June 8. LexisNexis Academic.

Nordhaus, William D. 2009. "Economic Issues in Designing a Global Agreement on Global Warming." Keynote address prepared for "Climate Change: Global Risks, Challenges, and Decisions," Copenhagen, Denmark.

Nossiter, Adam. 2010. "Far From Gulf, a Spill Scourge 5 Decades Old." *New York Times*, June 17. http://www.nytimes.com/2010/06/17/world/africa/17nigeria.html.

NRC (National Research Council). 2008. *Minerals, Critical Minerals, and the U.S. Economy*. Washington, DC: National Academies Press.

NRDC (National Resources Defense Council). 2011. "Good Wood: How Forest Certification Helps the Environment." January 13. http://www.nrdc.org/land/forests/qcert.asp.

Obach, Brian K. 2004. "New Labor: Slowing the Treadmill of Production?" *Organization & Environment* 17:337–354.

O'Connor, James. 1996. "The Second Contradiction of Capitalism." In *The Greening of Marxism*, edited by Ted Benton, 197–221. New York: Guilford.

———. 1998. *Natural Causes: Essays in Ecological Marxism*. New York: Guilford.

Oliveira-Souza, Henrique Freire de. 2000. "Genetically Modified Plants: A Need for International Regulation." *Annual Survey of International & Comparative Law* 6:129–174.

Otto, James, Craig Andrews, Fred Cawood, Michael Doggett, Pietro Guj, Frank Stermole, John Stermole, and John Tilton. 2006. "Mining Royalties: A Global Study of Their Impact on Investors, Government, and Civil Society." Washington, DC: World Bank.

Paliano, Guido Colonna di, Philip H. Trezise, and Nobuhiko Ushiba. 1977. "Directions for World Trade in the Nineteen-Seventies (1974)." In *Trilateral Commission Task Force Reports: 1–7*. New York: NYU Press.

Palmer, Mark. 2002. "Oil and the Bush Administration." *Earth Island Journal*, Autumn. http://thirdworldtraveler.com/Oil_watch/Oil_BushAdmin.html.

Park, Timothy, Mary Ahearn, Ted Covey, Kenneth Erickson, J. Michael Harris, Jennifer Ifft, Chris McGath, Mitch Morehart, Stephen Vogel, Jeremy Weber, and Robert Williams. 2011. "Agricultural Income and Finance Outlook." Washington, DC: U.S. Department of Agriculture.

Pascual, Francisco, and Arze Glipo. 2002. "WTO and Philippine Agriculture: Seven Years of Unbridled Trade Liberalization and Misery for Small Farmers." Institute for Agriculture and Trade Policy (IATP), January 28. http://www.iatp.org/documents/wto-and-philippine-agriculture-seven-years-of-unbridled-trade-liberalization-and-misery-fo.

Pateman, Carole. 1970. *Participation and Democratic Theory*. New York: Cambridge University Press.

Paul, Helena, Ricarda Steinbrecher, Devlin Kuyek, and Lucy Michaels. 2003. *Hungry Corporations: How Transnational Biotech Companies Colonise the Food Chain.* New York: Zed Books.

Peet, Richard. 2003. *Unholy Trinity: The IMF, World Bank, and WTO.* New York: Zed Books.

Peluso, Nancy Lee, and Michael Watts. 2001. *Violent Environments.* Ithaca: Cornell University Press.

Penney, Sarah, Jacob Bell, and John Balbus. 2009. "Estimating the Health Impacts of Coal-Fired Power Plants Receiving International Financing." New York: Environmental Defense Fund.

People for the American Way. 1996. *Buying a Movement: Right-Wing Foundations and American Politics.* Washington, DC: People for the American Way.

Perlez, Jane. 2006a. "Bow and Arrow Attack over U.S. Mining Company." *New York Times,* March 15. LexisNexis Academic.

———. 2006b. "Mining Protest Leaves 4 Dead in Indonesia." *New York Times,* March 17. LexisNexis Academic.

———. 2006c. "The Papuans Say, This Land and Its Ores Are Ours." *New York Times,* April 5. LexisNexis Academic.

Perlez, Jane, and Raymond Bonner. 2005. "Below a Mountain of Wealth, a River of Waste." *New York Times,* December 27. LexisNexis Academic.

Peschek, Joseph. 1987. *Policy Planning Organizations: Elite Agendas and America's Rightward Turn.* Philadelphia: Temple University Press.

Peterson, Peter G. 1971. "The United States in the Changing World Economy." Washington, DC: U.S. Government Printing Office.

Petty, Martin. 2011. "Pressure Mounts to Delay 'Dangerous' $3.5 Bln Mekong River Dam." Reuters, April 19. http://www.reuters.com/article/2011/04/19/laos-dam-idUSL3E7FI0YS20110419.

Pew Commission on Industrial Farm Animal Production. 2008. "Putting Meat on the Table: Industrial Farm Animal Production in America." Baltimore: Pew Commission on Industrial Farm Animal Production.

Pigato, Miria. 2000. "Foreign Direct Investment in Africa: Old Tales and New Evidence." Washington, DC: World Bank.

Pilger, John. 2000. "Labour Claims Its Actions Are Lawful While It Bombs Iraq, Starves Its People and Sells Arms to Corrupt States." JohnPilger.com, August 7. http://johnpilger.com/articles/labour-claims-its-actions-are-lawful-while-it-bombs-iraq-strarves-its-people-and-sells-arms-to-corrupt-states.

Pimentel, David. 2005. "Environmental and Economic Costs of the Application of Pesticides Primarily in the United States." *Environment, Development, and Sustainability* 7:229–252.

———. 2006. "Soil Erosion: A Food and Environmental Threat." *Environment, Development, and Sustainability* 8:119–137.

———. 2009. "Pesticides and Pest Control." In *Integrated Pest Management: Innovation-Development Process,* vol. 1, edited by Rajinder Peshin and Ashok K. Dhawan, 83–88. New York: Springer.

PNAC (Project for the New American Century). 2008. Home page. http://www. newamericancentury.org/.

Pollan, Michael. 2003. "The (Agri)Cultural Contradictions of Obesity." *New York Times*, October 12. http://www.nytimes.com/2003/10/12/magazine/12WWLN. html?pagewanted=all.

Porter, Ian. 2007. "The World Bank Group in the Greater Mekong Sub-region." Power-Point presentation at the Institute of Diplomacy and International Studies (IDIS), Rangsit University.

Poulgrain, Greg. 2005. "Indonesian Military Accused over Logging." *Courier Mail*, February 19. LexisNexis Academic.

Press Association. 2012. "BP Oil Spill Seriously Harmed Deep-Sea Corals, Scientists Warn." *Guardian*, March 26. http://www.guardian.co.uk/environment/2012/mar/26/ bp-oil-spill-deepwater-horizon.

Project Censored. 2004. *Censored 2004: The Top 25 Censored Stories of 2002–2003*. New York City: Seven Stories.

———. 2011. *Censored 2004: The Top 25 Censored News Stories of 2009–2010*. New York City: Seven Stories.

Public Citizen. 2012. "Fatally Flawed WTO Dispute System." Washington, DC: Public Citizen.

———. 2014. "Table of Foreign Investor-State Cases and Claims under NAFTA and Other U.S. Trade Deals." Washington, DC: Public Citizen.

Puzzanghera, Jim. 2001. "Bush Administration Oil Industry Ties." *San Jose Mercury News*, February 27. http://greenyes.grrn.org/2001/02/msg00097.html.

Quigley, Bill. 2008. "30 Years Ago Haiti Grew All the Rice It Needed. What Happened? The U.S. Role in Haiti's Food Riots." *Counterpunch*, April 21. http://www.counter-punch.com/quigley04212008.html.

Rainforest Action Network. 2012. "Chevron's Toxic Legacy in Ecuador." http://ran.org/ chevrons-toxic-legacy-ecuador (accessed October 5, 2012).

Ransley, Carol, Jonathan Cornford, and Jessica Rosien. 2008. "A Citizen's Guide to the Greater Mekong Subregion: Understanding the GMS Program and the Role of the Asian Development Bank." Victoria, Australia: Oxfam Australia.

Rastello, Sandrine. 2010. "U.S. Blocks IMF Board Proposal, Seeks Emerging Countries' Voice." Bloomberg Businessweek, August 20.

Reed, David. 1996. *Structural Adjustment, the Environment, and Sustainable Development*. London: Earthscan.

Reisch, Nikki, and Steve Kretzmann. 2008. "A Climate of War: The War in Iraq and Global Warming." Washington, DC: Oil Change International.

Renwick, Alan, Md. Mofakkarul Islam, and Steven Thomson. 2012. "Power in Agriculture: Resources, Economics and Politics." Lancashire, UK: Oxford Farming Conference.

Rice, James. 2009. "The Transnational Organization of Production and Uneven Environmental Degradation and Change in the World Economy." *International Journal of Comparative Sociology* 50:215–236.

Rice, Ronald E., ed. 2008. *Media Ownership: Research and Regulation.* Cresskill, NJ: Hampton.

Rich, Bruce. 1994. *Mortgaging the Earth: The World Bank, Environmental Impoverishment, and the Crisis of Development.* Boston: Beacon.

———. 2009. "Foreclosing the Future: Coal, Climate and Public International Finance." New York: Environmental Defense Fund.

Rieffel, Lex. 2003. *Restructuring Sovereign Debt: The Case for Ad Hoc Machinery.* Washington, DC: Brookings Institution Press.

Rio Tinto. 2004. "Rio Tinto Reaches Agreement to Sell Shares in FCX." Media release. March 22. http://www.riotinto.com/media/18435_media_releases_3383.asp.

———. 2013. "Rio Tinto 2013 Annual Report." London: Rio Tinto.

Roberts, J. Timmons, and Bradley C. Parks. 2007. *A Climate of Injustice: Global Inequality, North-South Politics, and Climate Policy.* Cambridge: MIT Press.

Roberts, J. Timmons, and Nikki Thanos. 2003. *Trouble in Paradise: Globalization and Environmental Crises in Latin America.* New York: Routledge.

Roberts, Jeremy, and Andrew Trounson. 2007. "Miner Accused of Evicting Townsfolk." *Australian,* July 3. LexisNexis Academic.

Robertson, Campbell, and Clifford Krauss. 2010. "Gulf Spill Is the Largest of Its Kind, Scientists Say." *New York Times,* August 2. http://www.nytimes.com/2010/08/03/us/03spill.html?_r=2&fta=y&pagewanted=print.

Robinson, Paul. 2005. "Colombian Pleads for BHP Help to Find Murderers: Wayuuan Villagers Want Justice after a Massacre of 12 People in a Colombian Mining Area." *Age,* May 14. LexisNexis Academic.

Rocha, Jan. 1986. "Mining Firms Rush into Amazon Indian Reserves." *Guardian,* May 3. LexisNexis Academic.

Rong, Feiwen, and Xia Yu. 2009. "Shortage of Rare Earths Used in Hybrids, TVs May Loom in China." Bloomberg News, September 3. http://www.bloomberg.com/apps/news?pid=newsarchive&sid=afn.hOk6pEHg.

Rose, Adam, and Thomas Tietenberg. 1993. "An International System of Tradable CO_2 Entitlements: Implications for Economic Development." *Journal of Environment and Development* 2:1–36.

Roseberry, William. 1996. "The Rise of Yuppie Coffees and the Reimagination of Class in the United States." *American Anthropologist* n.s. 98:762–775.

Ross, Michael. 2010. "Latin America's Missing Oil Wars." Washington, DC: World Bank Office of the Chief Economist for Latin America and the Caribbean.

Rotkin-Ellman, Miriam, Karen K. Wong, and Gina M. Solomon. 2012. "Seafood Contamination after the BP Gulf Oil Spill and Risks to Vulnerable Populations: A Critique of the FDA Risk Assessment." *Environmental Health Perspectives* 120:157–161.

Rubinson, Joel. 2009. "Empirical Evidence of TV Advertising Effectiveness." *Journal of Advertising Research* 49:220–226.

Russell, Grahame 2006. "Canadian Mining, the Mayan-Q'eqchi' People and the Cycles of Landlessness, Poverty and Repression." Upside Down World, December 6. http://www.upsidedownworld.org/main/content/view/534/1/.

Sainath, P. 2009. "The Largest Wave of Suicides in History." *Counterpunch*, February 12. http://www.counterpunch.org/2009/02/12/the-largest-wave-of-suicides-in-history/.

Salazar, Milagros. 2008. "Government Generosity Swells Mining Company Profits." Inter Press Service News Agency, February 4. http://ipsnews.net/news.asp?idnews=41061.

SAPRIN. 2004. *Structural Adjustment: The SAPRI Report (The Policy Roots of Economic Crisis, Poverty, and Inequality)*. New York: Zed Books.

Scahill, Jeremy. 2002. "No Fly Zones over Iraq." *Counterpunch*, December 4. http://www.counterpunch.org/2002/12/04/no-fly-zones-over-iraq/.

Scandinavian Oil Gas Magazine. 2010. "Chevron's $27 Billion Liability in Ecuador 'Glaringly Low' in Light of BP Disaster." June 10. http://www.scandoil.com/moxie-bm2/news/chevrons-27-billion-liability-in-ecuador-glaringly.shtml.

Schnaiberg, Allan. 1980. *The Environment*. New York: Oxford University Press.

Schnaiberg, Allan, and Kenneth Gould. 2000. *Environment and Society: The Enduring Conflict*. West Caldwell, NJ: Blackburn.

Schneider, Lambert. 2007. "Is the CDM Fulfilling Its Environmental and Sustainable Development Objectives? An Evaluation of the CDM and Options for Improvement." Freiburg, Germany: Öko-Institut.

Schoonover, Heather, and Mark Muller. 2006. "Food without Thought: How U.S. Farm Policy Contributes to Obesity." Minneapolis, MN: Institute for Agriculture and Trade Policy.

Scott, Rebecca. 2010. *Removing Mountains: Extracting Nature and Identity in the Appalachian Coalfields*. Minneapolis: University of Minnesota Press.

Sell, Susan 2003. *Private Power, Public Law: The Globalization of Intellectual Property Rights*. Cambridge: Cambridge University Press.

Shah, Anup. 2004. "The Bush Doctrine of Pre-emptive Strikes: A Global Pax Americana." Global Issues, April 24. http://www.globalissues.org/article/450/the-bush-doctrine-of-pre-emptive-strikes-a-global-pax-americana.

———. 2010a. "Poverty Facts and Stats." Global Issues, September 15. http://www.globalissues.org/article/26/poverty-facts-and-stats.

———. 2010b. "Today, Over 24,000 Children Died around the World." Global Issues, September 14. http://www.globalissues.org/article/715/today-over-24000-children-died-around-the-world.

Shandra, John M., Eran Shor, and Bruce London. 2008. "Debt, Structural Adjustment, and Organic Water Pollution: A Cross-National Analysis." *Organization & Environment* 21:38–55.

Shapiro, Robert J. 2009. "Is Cap and Trade a Dead Policy Walking?" *NDN* (blog), April 1. http://ndn.org/blog/2009/04/cap-and-trade-dead-policy-walking.

Shaw, Andrew. 1990. "CRVD's Carajas Project." *Mining Magazine*, August. LexisNexis Academic.

Shefrin, Hersh. 2002. *Beyond Greed and Fear: Understanding Behavioral Finance and the Psychology of Investing*. New York: Oxford University Press.

Shiva, Vandana. 1991. *The Violence of the Green Revolution*. London: Zed Book.

Silverstein, Ken. 2002. "U.S. Oil Politics in the 'Kuwait of Africa.'" *Nation*, April 22 (accessed July 10, 2010).

Simon, Herbert A. 1957. *Administrative Behavior: A Study of Decision-Making Processes in Administrative Organization*. New York: Macmillan.

———. 1978. "Rationality as Process and as Product of Thought." *American Economic Review* 68:1–16.

Singh, Bhoopendra, and Mudit Kumar Gupta. 2009. "Pattern of Use of Personal Protective Equipments and Measures during Application of Pesticides by Agricultural Workers in a Rural Area of Ahmednagar District, India." *Indian Journal of Occupation and Environmental Medicine* 13:127–130.

Sklar, Holly. 1980. *Trilateralism: The Trilateral Commission and Elite Planning for World Management*. Boston: South End.

Slovic, Paul. 1995. "The Construction of Preference." *American Psychologist* 50:364–371.

Smith, Stephen. 2011. *Environmental Economics: A Very Short Introduction*. New York: Oxford University Press.

Solly, Richard. 2010. "Colombia: Cerrejon Coal Mine." BHP Billiton Watch, November 18. http://bhpbillitonwatch.net/2010/11/18/cerrejon201/.

Soutar, Lindsay. 2009. "Unraveling the Greater Mekong Subregion Program: An Overview and Update on Key Structures, Programs and Developments." Victoria, Australia: Oxfam Australia.

Speth, James Gustave. 2005. *Red Sky at Morning: America and the Crisis of the Global Environment*. New Haven: Yale University Press.

St. Clair, Jeffrey, and Joshua Frank. 2008. "The Pentagon's Toxic Legacy." *Counterpunch*, May 12. http://www.counterpunch.org/2008/05/12/the-pentagon-s-toxic-legacy/.

Steger, Manfred B. 2009. *Globalization: A Very Short Introduction*. New York City: Oxford University Press.

Steinfeld, Henning, Pierre Gerber, Tom Wassenaar, Vincent Castel, Mauricio Rosales, and Cees de Haan. 2006. "Livestock's Long Shadow: Environmental Issues and Options." Rome: Food and Agriculture Organization of the United Nations.

Stenhouse, Renae. 2010. "Lao PDR Development Report 2010: Natural Resource Management for Sustainable Development." Washington, DC: World Bank.

Stiglitz, Joseph. 2002. *Globalization and Its Discontents*. New York City: Norton.

Stiles, Kendall. 1991. *Negotiating Debt: The IMF Lending Process*. Boulder, CO: Westview.

St. Louis Business Journal. 2006. "Monsanto Joint Venture Gets Approval for Corn." February 6. http://www.bizjournals.com/stlouis/stories/2006/02/06/daily9.html.

Stoff, Michael. 1980. *Oil, War, and American Security: The Search for a National Policy on Foreign Oil, 1941–1947*. New Haven: Yale University Press.

Stone, Randall. 2002. *Lending Credibility: The International Monetary Fund and the Post-Communist Transition*. Princeton: Princeton University Press.

———. 2004. "Lending Credibility in Africa." *American Political Science Review* 98:577–591.

Stueck, Wendy. 2006. "Sanctions Affect Ivanhoe's Myanmar Operation." *Globe and Mail*, April 6. LexisNexis Academic.

Suarez, Carlos, and Nicolas Rubio. 2011. "Haiti: Rice Production and Trade Update." Washington, DC: USDA Foreign Agricultural Service.

Survival International. 2000. "Uncontacted Indians Face Extinction." June 30. http://www.survival-international.org/news/121.

Tallontire, Anne, and Bill Vorley. 2005. "Achieving Fairness in Trading between Supermarkets and Their Agrifood Supply Chains." London: UK Food Group.

Taylor, C. Robert, and David Domina. 2010. "Restoring Economic Health to Contract Poultry Production." Report prepared for the Joint U.S. Department of Justice and U.S. Department of Agriculture / GIPSA Public Workshop on Competition Issues in the Poultry Industry, May 21, Normal, AL.

TerraChoice. 2009. "The Seven Sins of Greenwashing: Environmental Claims in Consumer Markets." London: TerraChoice.

Thampapillai, Dodo. 2002. *Environmental Economics: Concepts, Methods, and Policies.* New York: Oxford University Press.

Tillotson, James E. 2004. "America's Obesity: Conflicting Public Policies, Industrial Economic Development, and Unintended Human Consequences." *Annual Review of Nutrition* 24:617–643.

Tobin, Dave. 2012. "Update of Military Casualties in Iraq/Afghanistan Wars." *Post-Standard*, August 18. http://www.syracuse.com/news/index.ssf/2012/08/update_of_military_casualties.html.

Toler, Deborah. 1999. "The Right's 'Race Desk.'" FAIR, March 1. http://fair.org/extra-online-articles/the-right8217s-race-desk/.

Toussaint, Eric. 2005. *Your Money [or] Your Life: The Tyranny of Global Finance.* Translated by V. B. Manus, R. Krishnan, and G. Roche. Chicago: Haymarket Books.

Ture, Norman. 1981. "The Department of the Treasury." In *Mandate for Leadership*, edited by Charles L. Heatherly, 647–694. Washington, DC: Heritage Foundation.

Turse, Nick. 2011. "The US Has Been Bombing Iraq since 1991 without Stopping—Until Now." Alternet, November 14. http://www.alternet.org/story/153042/the_us_has_been_bombing_iraq_since_1991_without_stopping--until_now?paging=off.

Tversky, Amos, Paul Slovic, and Daniel Kahneman. 1990. "The Causes of Preference Reversal." *American Economic Review* 80:204–217.

Tyler, Patrick E. 1992. "U.S. Strategy Plan Calls for Insuring No Rivals Develop." *New York Times*, March 8, A1.

UNCTAD (United Nations Conference on Trade and Development). 2000. "The Least Developed Countries 2000 Report." New York: UNCTAD.

———. 2005. "Economic Development in Africa: Rethinking the Role of Foreign Direct Investment." New York: UNCTAD.

UNHCR (United Nations High Commissioner for Refugees). 2012. "2012 UNHCR Country Operations Profile—Afghanistan." http://www.unhcr.org/pages/49e486eb6.html.

United Nations. 2002. "Final Report of the Panel of Experts on the Illegal Exploitation of Natural Resources and Other Forms of Wealth of the Democratic Republic of the Congo." New York: United Nations.

United Nations Office of the High Commissioner for Human Rights. 2012. "Afghanistan Annual Report 2011: Protection of Civilians in Armed Conflict." Kabul, Afghanistan: United Nations Office of the High Commissioner for Human Rights.

U.S. Census Bureau. 2007. "Manufacturing: Subject Series: Concentration Ratios: Share of Value of Shipments Accounted for by the 4, 8, 20, and 50 Largest Companies for Industries: 2007 Economic Census." Washington, DC: U.S. Census Bureau.

USDA (U.S. Department of Agriculture). 2013. "Genetically Engineered Varieties of Corn, Upland Cotton, and Soybeans, by State and for the United States, 2000–13." Washington, DC: USDA. http://www.ers.usda.gov/data-products/adoption-of-genetically-engineered-crops-in-the-us.aspx#.U58PkCjLIis.

USDA (U.S. Department of Agriculture) Foreign Agricultural Service. 2015. "Exporting." Washington, DC: USDA Foreign Agricultural Service. http://www.fas.usda.gov/topics/exporting (accessed March 31, 2015).

U.S. Department of Defense. 2012. "Casualty Fact List." Washington, DC: U.S. Department of Defense. http://www.defense.gov/news/casualty.pdf (accessed October 24, 2012).

U.S. Department of State. 1999. "Burkina Faso: US Department of State Country Reports on Human Rights Practices." Washington, DC: U.S. State Department. http://www.state.gov/g/drl/rls/hrrpt/1999/229.htm.

———. 2006a. "Burma: Country Reports on Human Rights Practices." Washington, DC: U.S. Department of State, March 8. http://www.state.gov/g/drl/rls/hrrpt/2005/61603.htm.

———. 2006b. "Gabon: Country Reports on Human Rights Practices." , Washington, DC: U.S. Department of State, March 8. http://www.state.gov/g/drl/rls/hrrpt/2005/61570.htm.

U.S. Department of Veterans Affairs. 2012. "Public Health: Burn Pits." Washington, DC: U.S. Department of Veterans Affairs. http://www.publichealth.va.gov/exposures/burnpits/index.asp (accessed October 19, 2012).

USGAO (U.S. Government Accountability Office). 2003. "Energy Task Force: Process Used to Develop the National Energy Policy." U.S. Washington, DC: USGAO.

———. 2010. "SUPERFUND: Interagency Agreements and Improved Project Management Needed to Achieve Cleanup Progress at Key Defense Installations." Washington, DC: USGAO.

USGS (U.S. Geological Survey). 2006. *Minerals Yearbook*. Washington, DC: USGS.

———. 2008. *Mineral Commodity Summary*. Washington, DC: USGS.

Vaisse, Justin. 2010. *Neoconservatism: The Biography of a Movement*. Translated by A. Goldhammer. Cambridge: Belknap Press of Harvard University Press.

Vallette, Jim, Daphne Wysham, and Nadia Martínez. 2004. "A Wrong Turn from Rio: The World Bank's Road to Climate Catastrophe." Washington, DC: Sustainable Energy and Economy Network, Institute for Policy Studies.

van den Putte, Bas. 2009. "What Matters Most in Advertising Campaigns? The Relative Effect of Media Expenditure and Message Content Strategy." *International Journal of Advertising* 28:669–690.

Van Natta, Don, and Neela Banerjee. 2002. "Top G.O.P. Donors in Energy Industry Met Cheney Panel." *New York Times*, A1, A19.

Victor, David. 2009. "Global Warming Policy after Kyoto: Rethinking Engagement with Developing Countries." Stanford: Freeman Spogli Institute for International Studies, Stanford University.

Vidal, John. 2007. "Sold Down the River." *Guardian*, September 22. LexisNexis Academic.

Vorley, Bill. 2004. "Food Inc.: Corporate Concentration from Farm to Consumer." London: UK Food Group.

Vostroknutova, Ekaterina. 2010. "Lao PDR Development Report 2010: Natural Resource Management for Sustainable Development, Hydropower and Mining." Washington, DC: World Bank.

Vreeland, James. 2005. "The International and Domestic Politics of IMF Programs." Working paper. New Haven: Department of Political Science, Yale University.

———. 2007. *The International Monetary Fund: Politics of Conditional Lending*. New York: Routledge.

Wallach, Lori, and Patrick Woodall. 2004. *Whose Trade Organization? A Comprehensive Guide to the WTO*. New York: New Press.

Wallerstein, Immanuel. 2003. *The Decline of American Power: The U.S. in a Chaotic World*. New York: New Press.

Walton, Abigail Abrash. 2001. "Mining a Sacred Land—Freeport in West Papua." Carnegie Council on Ethics and International Affairs. Mines and Communities, April 23. http://www.minesandcommunities.org/article.php?a=1009.

Warner, Jerry, and P. W. Singer. 2009. "Fueling the 'Balance': A Defense Energy Strategy Primer." Washington, DC: Brookings Institution.

Watts, Michael. 2008. "Economies of Violence: More Oil, More Blood." In *Contested Grounds: Essays on Nature, Culture, and Power*, edited by Amita Baviskar, 106–136. New York: Oxford University Press.

Weise, Elizabeth. 2011. "More of World's Crops Are Genetically Engineered." *USA Today*, February 22. http://www.usatoday.com/tech/news/biotech/2011-02-22-biotech-crops_N.htm#.

Wessel, David. 2009. "Pollution Politics and the Climate-Bill Giveaway." *Wall Street Journal*, May. http://online.wsj.com/article/SB124304449649349403.html.

White, T. Kirk, and Robert A. Hoppe. 2012. "Changing Farm Structure and the Distribution of Farm Payments and Federal Crop Insurance." Washington, DC: U.S. Department of Agriculture.

White House Oil Spill Commission. 2010. "America's Gulf Coast: A Long Term Recovery Plan after the Deepwater Horizon Oil Spill." Washington, DC: White House Oil Spill Commission.

Wiesmeth, Hans. 2012. *Environmental Economics: Theory and Policy in Equilibrium*. New York: Springer-Verlag.

Williams, Albert L. 1971. "United States International Economic Policy in an Interdependent World." Washington, DC: U.S. Government Printing Office.

Williamson, Oliver E. 1981. "The Economics of Organization: The Transaction Cost Approach." *American Journal of Sociology* 87:548–577.

Wise, Timothy. 2005. "Identifying the Real Winners from U.S. Agricultural Policies." Medford, MA: Global Development and Environment Institute, Tufts University.

———. 2011. "Mexico: The Cost of U.S. Dumping." *NACLA* 44:47–48.

Wise, Timothy, and Betsy Rakocy. 2010. "Hogging the Gains from Trade: The Real Winners from U.S. Trade and Agricultural Policies." Medford, MA: Global Development and Environment Institute, Tufts University.

Woods, Ngaire. 2000. "The Challenges of Multilateralism and Governance." In *The World Bank: Structure and Policies,* edited by Christopher L. Gilbert and David Vines, 132–152. New York: Cambridge University Press.

———. 2001. "Making the IMF and the World Bank More Accountable." *International Affairs* 77:83–100.

———. 2006. *The Globalizers: The IMF, the World Bank, and Their Borrowers.* Ithaca: Cornell University Press.

World Bank. 1982. "Report and Recommendation of the President of the International Bank for Reconstruction and Development to the Executive Directors on a Proposed Loan in an Amount of US$304.5 Million to Companhia Vale Do Rio Doce with the Guarantee of the Federative Republic of Brazil for the Carazas Iron Ore Project." Washington, DC: World Bank.

———. 1994. "Resettlement and Development: The Bankwide Review of Projects Involving Involuntary Resettlement 1986–1993." Washington, DC: World Bank.

———. 1997. "Adjustment Lending in Sub-Saharan Africa: An Update." Washington, DC: World Bank.

———. 2007. "Strategy Note on World Bank Regional Support for the Greater Mekong Sub-Region." Washington, DC: World Bank Southeast Asia Country Management Unit, East Asia and Pacific Region.

———. 2010. "Lao PDR: Project Overview and Description." Washington, DC: World Bank, September. http://web.worldbank.org/WBSITE/EXTERNAL/COUNTRIES/EASTASIAPACIFICEXT/LAOPRDEXTN/0,,contentMDK:21109109pagePK:141137piPK:141127theSitePK:293684,00.html.

———. 2011a. "Boards of Directors." Washington, DC: World Bank. http://web.worldbank.org/WBSITE/EXTERNAL/EXTABOUTUS/ORGANIZATION/BODEXT/0,,pagePK:64020055theSitePK:278036,00.html (accessed March 11, 2011).

———. 2012. "Boards of Governors." Washington, DC: World Bank. http://web.worldbank.org/WBSITE/EXTERNAL/EXTABOUTUS/0,,contentMDK:20873632menuPK:1696997pagePK:51123644piPK:329829theSitePK:29708,00.html (accessed August 15, 2013).

———. n.d. "Summary of Report on Adjustment in Sub-Saharan Africa." Washington, DC: World Bank. http://lnweb90.worldbank.org/oed/oeddoclib.nsf/b57456d58aba40e585256ad400736404/0e5bd1800024b2ce852567f5005d8617?OpenDocument (accessed May 27, 2011).

World Commission on Dams. 2000. *Dams and Development: A New Framework for Decision-Making.* London: Earthscan.

Wright, Erik Olin. 2010. *Envisioning Real Utopias.* New York: Verso.

Xi, Li. 2006. "Thousand Armed Police Quell Protest against Manganese Processing Plant." *Epoch Times,* July 18. LexisNexis Academic.

Xu, L. F, and M. L. Liu. 1999. "Impacts of Rare-Earth Mining on the Environment and the Effects of Ecological Recovery Measures on Soils." In *Remediation and Management of Degraded Lands,* edited by M. H. Wong, J. W. C. Wong, and A. J. M. Baker, 103–110. Boca Raton, FL: CRC.

Yergin, Daniel. 1991. *The Prize: The Epic Quest for Oil, Money, and Power.* New York: Simon and Schuster.

Yi-Chong, Xu, and Patrick Weller. 2009. *Inside the World Bank: Exploding the Myth of the Monolithic Bank.* New York: Palgrave Macmillan.

York, Richard. 2004. "The Treadmill of (Diversifying) Production." *Organization & Environment* 17:355–362.

———. 2008. "De-carbonization in Former Soviet Republics, 1992–2000: The Ecological Consequences of De-modernization." *Social Problems* 55:370–390.

York, Richard, and Eugene A. Rosa. 2003. "Key Challenges to Ecological Modernization Theory." *Organization & Environment* 16:273–288.

Zhou, Nan, Dongsheng Zhou, and Ming Ouyang. 2003. "Long-Term Effects of Television Advertising on Sales of Consumer Durables and Nondurables: The Case of China." *Journal of Advertising* 32:45–54.

Ziegler, Jean. 2006. "The Right to Food: Report of the Special Rapporteur on the Right to Food; Addendum: Mission to Guatemala." New York: United Nations Economic and Social Council.

INDEX

Abrams, Elliot, 245

ACTN. *See* Advisory Committee on Trade Negotiations

ADB. *See* Asian Development Bank

ADM. *See* Archer Daniels Midland

Advertising: consumer behavior with, 22–23; environmental claims in, 23–24

Advisory Committee on Trade Negotiations (ACTN), 161, 162

AEI. *See* American Enterprise Institute

Afghanistan, 276n21; death toll in, 235–36; toxic waste and, 233–34; U.S. war in, 230, 231, 257

Africa: mining projects in, 13, 167, 172, 192–93, 274n10; mining revenues in, 204–5, 274n10; poverty in, 73; produce exports in, 137–38; structural adjustment in, 68, 70, 71–72. *See also* Sub-Saharan Africa

African Command (Africom), 200, 231

Agreement on Agriculture (AoA), 56, 116, 147; creation of, 151–52, 153, 272n18; developing nation farmers and, 157–59; subsidies in, 272n17; trade liberalization and, 157

Agreement on Trade-Related Aspects of Intellectual Property Rights (TRIPs), 117, 147–48; developing nations and, 160–61; elite-controlled organizations and, 161–62; outline of, 159–60

Agribusiness firms: contract use in, 128–32; farmer choices set by, 122, 139; international treaties and, 148; joint ventures in, 126, 128; largest U.S., 126, *127*, 128, *128*; market control of, 115,

125, 126, 130; trade policy and, 12, 116, 155–57

Agricultural commodity chains: elites use of, 55, 56, 116, 123; environmental crisis and, 6–7; oligopoly and oligopsony power in, 12, 116, 123–28, 139, 145–47, 163; vertically structured, 117–18. *See also* Captive value chains

Agricultural trade policy, U.S.: AEI and, 156; policy planning network and, 116; Trilateral Commission and, 157; WTO and, 12, 116

Agriculture, modern, 1, 11, 163; capital accumulation and, 116, 148, 162, 165; climate change and, 115, 120–21; crop diversity in, 121; elite-control of, 12, 116, 151; farmer hardship in, 115, 122, 139; fossil fuel dependence in, 115, 117, 120; greenhouse gas emissions with, 120–21; health risks in, 118–19, 132; IDE model and, 135; industry in U.S., 116, 122, 125; market competition in, 141, 142; market concentration in, 125–26; market manipulation in, 128–32; pesticides in, 118–19, 271n2; productivity of, 121, 271n1; property rights in, 116; social and environmental harm with, 116, 117–23; soil erosion in, 71, 72, 115, 119; structural adjustment on, 71; technology in, 118, 121–22, 139; trade liberalization in, 12, 116, 148–51; water use in, 119–20; after World War II, 115, 118, 130–31. *See also* Farmers; Monoculture farming; Seed market

313

ABOUT THE AUTHOR

Liam Downey is Associate Professor of Sociology and Faculty Associate for Environmental Studies at the University of Colorado at Boulder. He is the author of numerous journal articles examining the environmental consequences of social, economic, and political inequality.